DMU 0653748 01 2

AF606171

DE MONTFORT UNIVERSITY LIBRARY

Department of
Library Services

Normal Loan
Kimberlin Library

De Montfort University www.**library**.dmu.ac.uk

Please return this item on or before the date indicated below (if you have used Self Service please write in the date yourself as a reminder – the due date is printed on your receipt).

Items are issued subject to reservation recall. See Library Guide for details. Fines will be charged for the late return or late renewal of items.

24 hour renewals can be made by telephoning **(0116) 257 7043**, or you can renew online via the Library Catalogue under 'My Account'.

General library information: (0116) 257 7042 (service hours)

24 JUN 2011

22 JUN 2012

Form 14 PC28

Evolving Rule-Based Models

Studies in Fuzziness and Soft Computing

Editor-in-chief
Prof. Janusz Kacprzyk
Systems Research Institute
Polish Academy of Sciences
ul. Newelska 6
01-447 Warsaw, Poland
E-mail: kacprzyk@ibspan.waw.pl
http://www.springer.de/cgi-bin/search_book.pl?series=2941

Further volumes of this series can be found at our homepage.

Vol. 72. M. Mareš
Fuzzy Cooperative Games, 2001
ISBN 3-7908-1392-3

Vol. 73. Y. Yoshida (Ed.)
Dynamical Aspects in Fuzzy Decision, 2001
ISBN 3-7908-1397-4

Vol. 74. H.-N. Teodorescu, L.C. Jain and A. Kandel (Eds.)
Hardware Implementation of Intelligent Systems, 2001
ISBN 3-7908-1399-0

Vol. 75. V. Loia and S. Sessa (Eds.)
Soft Computing Agents, 2001
ISBN 3-7908-1404-0

Vol. 76. D. Ruan, J. Kacprzyk and M. Fedrizzi (Eds.)
Soft Computing for Risk Evaluation and Management, 2001
ISBN 3-7908-1406-7

Vol. 77. W. Liu
Propositional, Probabilistic and Evidential Reasoning, 2001
ISBN 3-7908-1414-8

Vol. 78. U. Seiffert and L.C. Jain (Eds.)
Self-Organizing Neural Networks, 2002
ISBN 3-7908-1417-2

Vol. 79. A. Osyczka
Evolutionary Algorithms for Single and Multicriteria Design Optimization, 2002
ISBN 3-7908-1418-0

Vol. 80. P. Wong, F. Aminzadeh and M. Nikravesh (Eds.)
Soft Computing for Reservoir Characterization and Modeling, 2002
ISBN 3-7908-1421-0

Vol. 81. V. Dimitrov and V. Korotkich (Eds.)
Fuzzy Logic, 2002
ISBN 3-7908-1425-3

Vol. 82. Ch. Carlsson and R. Fullér
Fuzzy Reasoning in Decision Making and Optimization, 2002
ISBN 3-7908-1428-8

Vol. 83. S. Barro and R. Marín (Eds.)
Fuzzy Logic in Medicine, 2002
ISBN 3-7908-1429-6

Vol. 84. L.C. Jain and J. Kacprzyk (Eds.)
New Learning Paradigms in Soft Computing, 2002
ISBN 3-7908-1436-9

Vol. 85. D. Rutkowska
Neuro-Fuzzy Architectures and Hybrid Learning, 2002
ISBN 3-7908-1438-5

Vol. 86. M.B. Gorzałczany
Computational Intelligence Systems and Applications, 2002
ISBN 3-7908-1439-3

Vol. 87. C. Bertoluzza, M.Á. Gil and D.A. Ralescu (Eds.)
Statistical Modeling, Analysis and Management of Fuzzy Data, 2002
ISBN 3-7908-1440-7

Vol. 88. R.P. Srivastava and T.J. Mock (Eds.)
Belief Functions in Business Decisions, 2002
ISBN 3-7908-1451-2

Vol. 89. B. Bouchon-Meunier, J. Gutiérrez-Ríos, L. Magdalena and R.R. Yager (Eds.)
Technologies for Constructing Intelligent Systems 1, 2002
ISBN 3-7908-1454-7

Vol. 90. B. Bouchon-Meunier, J. Gutiérrez-Ríos, L. Magdalena and R.R. Yager (Eds.)
Technologies for Constructing Intelligent Systems 2, 2002
ISBN 3-7908-1455-5

Vol. 91. J.J. Buckley, E. Eslami and T. Feuring
Fuzzy Mathematics in Economics and Engineering, 2002
ISBN 3-7908-1456-3

Plamen P. Angelov

Evolving Rule-Based Models

A Tool for Design of Flexible Adaptive Systems

With 106 Figures
and 9 Tables

Physica-Verlag
A Springer-Verlag Company

Dr. Plamen P. Angelov
Loughborough University
Department of Civil and Building Engineering
Loughborough, LE11 3TU
United Kingdom
p.p.angelov@lboro.ac.uk

DE MONTFORT UNIVERSITY
LIBRARY:
B/code
Fund: 36 Date: 19/05/10
Sequence:
Class: 003.7
Suffix: ANG

ISSN 1434-9922
ISBN 3-7908-1457-1 Physica-Verlag Heidelberg New York

Cataloging-in-Publication Data applied for
Die Deutsche Bibliothek – CIP-Einheitsaufnahme
Angelov, Plamen P.: Evolving rule based models: a tool for design of flexible adaptive systems; with 9 tables / Plamen P. Angelov. – Heidelberg; New York: Physica-Verl., 2002
(Studies in fuzziness and soft computing; Vol. 92)
ISBN 3-7908-1457-1

This work is subject to copyright. All rights are reserved, whether the whole or part of the material is concerned, specifically the rights of translation, reprinting, reuse of illustrations, recitation, broadcasting, reproduction on microfilm or in any other way, and storage in data banks. Duplication of this publication or parts thereof is permitted only under the provisions of the German Copyright Law of September 9, 1965, in its current version, and permission for use must always be obtained from Physica-Verlag. Violations are liable for prosecution under the German Copyright Law.

Physica-Verlag Heidelberg New York
a member of BertelsmannSpringer Science+Business Media GmbH

© Physica-Verlag Heidelberg 2002
Printed in Germany

The use of general descriptive names, registered names, trademarks, etc. in this publication does not imply, even in the absence of a specific statement, that such names are exempt from the relevant protective laws and regulations and therefore free for general use.

Hardcover Design: Erich Kirchner, Heidelberg

SPIN 10858015 88/2202-5 4 3 2 1 0 – Printed on acid-free paper

To Rossi, Lachko, Mariela,
and my parents

PREFACE

The idea about this book has evolved during the process of its preparation as some of the results have been achieved in parallel with its writing. One reason for this is that in this area of research results are very quickly updated. Another is, possibly, that a strong, unchallenged theoretical basis in this field still does not fully exist.

From other hand, the rate of innovation, competition and demand from different branches of industry (from biotech industry to civil and building engineering, from market forecasting to civil aviation, from robotics to emerging *e*-commerce) is increasingly pressing for more ***customised*** solutions based on learning consumers behaviour. A highly interdisciplinary and rapidly innovating field is forming which focus is the design of ***intelligent*, self-adapting** systems and machines. It is on the crossroads of control theory, artificial and computational intelligence, different engineering disciplines borrowing heavily from the biology and life sciences. It is often called *intelligent* control, *soft computing* or *intelligent* technology.

Some other branches have appeared recently like *intelligent* agents (which migrated from robotics to different engineering fields), data fusion, knowledge extraction etc., which are inherently related to this field. The core is the attempts to enhance the abilities of the *classical* control theory in order to have more adequate, *flexible,* and adaptive models and control algorithms.

The ability **to learn new rules, to improve system behaviour** through learning, **preserving** in the same time the useful **previous experience** is the trademark of intelligence and thus the ultimate challenge to *intelligent* control. It offers possibilities to design truly *intelligent* machines, constantly to extract and update knowledge from the processes of different fields.

The aim of this book is to present in a systematic way the ***e***volving **r**ule-based (***e*R**) models as a powerful tool for ***adaptive flexible*** description of systems. They could find application in fields like behaviour modelling and *customised* systems design (e.g. customised comfort requirements), fault detection and diagnostics, market forecasting, risk assessment, adaptive model-based control, *intelligent* agents etc.

The book represents the results of the recent work of the author and his accumulated experience during last decade in the field of modelling, optimisation, and control using so called *intelligent techniques* (*fuzzy* set theory, genetic algorithms, neural networks, data mining, and knowledge extraction). Engineering applications in different fields (in building services engineering, biotech industry, and risk assessment) are considered, which has been developed in the framework of research projects, in which the author participated recently.

The author would like kindly to thank his colleagues from the Building Services Research Group of the Department of Civil and Building, Loughborough, especially Dr. Richard Buswell and Dr. Jonathan Wright with whom he collaborated actively recently, Prof. Victor Hanby and Prof. Dennis Loveday.

Special thanks are addressed to Dr. Dimitar Filev, the Ph.D. supervisor of the author, who introduced him in the world of science and fuzzy set theory more than a dozen years ago, who has significant role in inspiring and forming his research interests. The author is very much indebted to Dr. Filev also for the useful comments on the manuscript and to Prof. Janusz Kacprzyk for the support in preparation of the manuscript and making it in a form most convenient and useful for the readers.

The author's gratitude is also for Dr. Reinhard Guthke, with whom the author collaborated in the field of biotech applications during his stays in Hans-Knoell Institute, Jena, Germany (in 1995 and in 1996) and to Dr. Dimiter Lakov, with whom the author collaborated on risk assessment. Collaboration and support of the following colleagues during the previous years is also appreciated: Prof. Gancho Vachkov, Dr. Ljudmila Kuncheva, Dr. Yordan Kostov, and Prof. Krassimir Atanassov.

Last, but not the least, this book would be impossible without constant and unwavering support, patience, love and the inspiration from Rossi, Lachko, and Mariela.

Loughborough, UK Plamen P. Angelov
July 2001

CONTENTS

1 INTRODUCTION

1.1
Flexible Models – an Opportunity for Control Theory

Control theory is nowadays a well developed and structured one, especially its linear part, including system identification (Ljung, 1987) and adaptive systems (Astrom and Wittenmark, 1984). Real engineering applications, however, very often does not comply with the rigorous assumptions on which this theory is based.

Until 1980s most commonly used have been so called *first principle* models, normally based on mass- and energy balance, and their linearised versions. They, however, are very often inadequate or practically difficult, even impossible, to be build (Driankov et. al., 1993; Yager and Filev, 1994). They consider many factors as a disturbance and normally ignore time variations. This makes them applicable only locally. Practically, significant amount of process knowledge is qualitative and imprecise, and, therefore is ignored (Babuska et. al., 1999).

The other type of models, used in practice, is the so-called *black-box* type (including polynomial, regression models, and, more recently, neural networks). Generally, they are an excellent tool for approximation: it has been theoretically proven that a three-layered neural network is able to approximate an arbitrary non-linear function (Hornik, 1991). But, they have the following basic shortfalls:

- they are not transparent (their coefficients are not related to the process characteristics; they did not explain *why* the object behaves in a certain way; they are not appropriate for inspection and analysis);
- they could not incorporate existing knowledge (it is almost impossible to add existing expert or common knowledge, except in formulating ranges of variables and parameters and determining the order of the model);
- they are data-based, non generic (they are data-dependent and are limited to the range of variables considered during the training);
- they require structure of the model (order, number of neurones etc.) to be *known a priori*.

An alternative approach to modelling systems has been developed by Mamdani and Assilian (1975), Pedrycz (1984), Takagi and Sugeno (1985). It is based on *fuzzy set* theory, introduced by Professor Lotfi Zadeh in his famous pioneering paper (Zadeh, 1965).

Initially *fuzzy* sets have been seen as a "pure" mathematical tool for generalised reasoning and logic. Its power has become more obvious only after it found successful engineering applications, mostly during the last decade. Nowadays, *fuzzy* models, *fuzzy* controllers and *fuzzy* systems could be found in many commercial products, ranging from *intelligent* decision support systems, *intelligent* search engines and robotic complexes to cameras, camcorders, cars and washing machines. Controllers, based on *fuzzy* logic are responsible for the Space Shuttle (Lea et. al., 1996).

Fuzzy logic is coming into our homes by *intelligent* appliances. Nevertheless, it is still very often to find scepticism of many people and even specialists, especially engineers educated some decades ago, in *fuzzy* theory and *fuzzy* technology. The possible reason is that people associate its meaning with the degree of the accuracy and even the seriousness of the system, especially if it is engineering one.

There is a degree of imprecision, inadequacy in the term *fuzzy* itself. The core element in the definition of a *fuzzy* set is the membership to it. In fact, a *fuzzy* set supposes **partial** membership to itself or **combined**, ***flexible*** membership to several close sets. Thus the structure of *fuzzy* sets is *flexible* in terms of membership to it. Therefore, terms ***flexible*** and ***fuzzy*** will be used in this book as synonyms in this context.

1.2 *Flexible* Models and their Identification

Two basic groups of *flexible* models have been introduced during the last two decades:

- *Flexible* rule-based (FRB) models
  - linguistic models with *flexible* inputs and *flexible* outputs (Mamdani, 1977);
  - *flexible* relational models (Pedrycz, 1984);
  - models with *flexible* inputs and precise outputs (Takagi and Sugeno, 1985).
- Models with *flexible* parameters or (in)equalities
  - models with *flexible* inequalities and equalities (Zimmermann, 1983);
  - models with *flexible* parameters (Tanaka and Asai, 1984);

The second group of models has application mainly in optimisation problems, including linear (Rommelfanger, 1999) and mathematical programming (Carlsson and Fuller, 2001), and optimal control (Filev and Angelov, 1992).

FRB models could effectively combine measured quantitative data with the operational experience and qualitative and imprecise information (Babuska et. al., 1999). They are transparent (by differ from neural networks, polynomial, regression and other *black-box* type of models) and they are able to express and incorporate human preferences, perceptions and subjectivity.

This *flexible* description and its structure and parameter identification methods had significant impact on *flexible* systems research in general (Dubois et. al., 1998):

- it suggested that these models could be used as a tool for modelling non-linear systems (in fact, practically all real systems are non-linear);
- it shows that these models are also universal approximators (Wang, 1992), preserving, however, its linguistic concept and explanatory potential (especially useful in fault detection and diagnostics, genome decoding, *intelligent* systems, agents, robotics etc.);
- it also suggested that these models could be designed based on objective data, but not from expert knowledge only.

FRB models of Takagi-Sugeno type have found wide application (including in process control), because of their computational and interpretation simplicity and efficiency. This type of models, often called TSK (from Takagi-Sugeno-Kang), will be considered later as a basis for the ***e***volving **R**ule-based (***e*R**) models, introduced in the book.

TSK models make interpolation, blending between several (normally linear) models of the same system, each being valid in a certain sub-space of states. In many engineering applications exist such sub-spaces, like "exponential phase", "lag phase", "saturation" etc. In TSK models these zones are described with *fuzzy* sets, considering partial and combined membership, making possible a smooth transition between neighbouring zones.

Practically, however, the main stumbling blocks in the design of FRB models is the generic, adequate, time wise and computation wise effective generation of their structure (rule base, membership functions, linguistic labels) and parameters.

1.2.1 Expert Knowledge and Parameters Tuning

One tendency, typical until early 1990's has been to rely basically on existing expert knowledge and only to tune *fuzzy* sets' parameters by linear least squares, gradient-based or genetic algorithms (Nozaki et. al., 1995; Furuhashi et. al., 1995).

This process of tuning, applied also to the *flexible* logic controllers, is often called optimisation of *flexible* models. It is analogous to the process of parameter identification of *conventional* models, which is also an optimisation problem itself. The main difference is that in this case the model structure is provided by experts.

Practically, very often extracting and proper formulation of the expert knowledge is a tedious and subjective procedure. It is an *ill defined* process itself and requires unpopular, time consuming and, often, low-effective tools like questionnaires, interviews and an always difficult working collaboration between experts in a domain and a control engineer, experienced in *fuzzy* set theory. This approach is also very much dependent on the level of expertise of the experts and their ability to express it in an understandable way.

A combination of this approach with the more objective data-driven technique, producing *hybrid* models seems a promising direction for future developments.

1.2.2
Data-driven Techniques

During the last few years, so-called *data-driven* techniques (Cios et. al., 1998) or *rule/knowledge extraction* has been intensively developed (Hoffmann and Pfister, 1996; Carse et. al., 1996). They are making an attempt to identify both the model structure and parameters based on data mainly (expert knowledge could also be used, but is not essential).

The recent "boom" in this area could be explained by the fact that presently huge amounts of raw data exist in literally every branch of human activity, while few decades ago it was a real problem to obtain, verify and validate the data. This fact is one of the consequences of the tremendous development of computer and information technology: nowadays it is not a problem to acquire, transform and transmit to any point of the globe huge amount of data. Currently, the real problem of many industries and companies is how effectively to cope with the exponentially growing data bases, how to take into account qualitative and imprecise information, which they could content, but not how to find them as it was very often few decades ago.

The approaches of this group, which just have appeared during the last few years, could basically be divided into two groups:

- **Quasi-linear**

  They treat separately the antecedent and consequent parts of the *flexible* rules and applies clustering and linear least squares approaches (Yager and Filev, 1993; Chiu, 1994);

- **Non-linear**

  Both structure of antecedent and consequent parts as well as parameters of the *flexible* model are determined numerically by a non-linear search algorithm, normally a gradient-based procedure or a genetic algorithm (Shimojima et. al., 1995; Castillo et. al., 2001).

Basic algorithms, which are applied for clustering are fuzzy C-means (Bezdek, 1974), which is iterative one and *subtractive clustering* (Chiu, 1994), which itself is an improved version of the so-called *mountain clustering* (Yager and Filev, 1993). The latest clustering algorithms are non-iterative and are used as a basis of the *on-line* approach, introduced in this book.

1.2.3 Precision and Transparency

The swift success of data-driven approaches and related developments in neural networks (the appearance of so called radial-basis functions, which are making an attempt to increase the transparency of this type of *black-box* models) lead to the discussion about the nature of *flexible* rule-based models (Dubois et. al., 1998). One extreme in this dispute is arguing that *flexible* models are best appropriate for representation of expert, human knowledge only, as this was their original purpose and field of application. Another extreme is to consider them as a *purely* numerical universal approximator (Wang, 1992).

The position of the author on this issue is that, while being able successfully to compete with other *purely black-box type* models, the huge advantage of the *flexible* models is their potential easy and in a natural way to incorporate existing expert knowledge, linguistic and mixed information or to be used as a source for extraction of an information understandable for human experts. Different schemes for hybrid models and use of mixed type of information exists and have been reported (Chen et. al., 2000).

Transparency of FRB models is, normally, balanced with the precision in modelling and the potential to explain its information content (Setnes and Rubos, 2000; Castillo et.al., 2001). In some applications the precision of the model is paramount and then it could be a better choice to use another type of models, for example neural networks.

In cases, when the structure of the model and its transparency is very important, like in fault detection and diagnosis, for example, neural networks alone are of little help. Interpretability of the *flexible* models could be further improved by simplification of their structure based on similarity.

In majority of practical problems, both satisfactory prediction and good generalisation properties combined with possibility to inspect and analyse it are highly required. Therefore, the conviction of the author is that *flexible* rule-based models are a promising option and current developments in this field, including their *on-line* identification, presented in this book, make them more and more attractive and efficient tool for practical use.

1.2.4
The Need for *On-line* Algorithms

All of the known methods for *flexible* rule-based models' identification, however, are directly applicable only in *off-line* mode. Their *on-line* application is possible for the price of re-training the whole structure and all parameters with iterative and time-consuming procedures, like *error back-propagation* or GA.

Some processes (building thermal systems, biotechnological processes etc.) has relatively slow dynamics making such re-training practically possible. In this way, however, the model structure has not been preserved and the search starts every time from a new starting point. In fact, these are procedures of repeated building new models. In the same time, new data could update or alter not necessarily the whole structure of the model, but rather a small part of it (one or few rules). This makes re-training of the whole model not necessary and not efficient.

The appearance of a new rule could indicate an area of the state space variables, which has not been covered by the initial training data, a new characteristic of the process or reaction to a new disturbance. In fact, many regimes and process states could not be practically included into the training data (like zero mass flow rates or pressure, faulty process behaviour etc.), but states close enough to them could well appear during the process run. Besides, every new data collected *on-line* posses additional describing potential, which have to be compared to that of the training data and, if effective, to be used for correction of existing or generation of a new rule.

The approach, treated in this book, drives the evolution of the rule base by identifying which rule to be replaced and generates a new one, if significant new data are collected (based on the descriptive potential of each data sample), and does it *on-line* in a ***non-iterative*** and ***recursive*** manner. In case the informative potential of the new data sample is high enough and it is not too close to an existing rule, it is added to the rule base without replacement of an old rule. Therefore, learning could start practically without *a priory* information and only few data samples, which makes the approach potentially very useful in robotics.

A significant problem in application of *flexible* models is that there is no convenient and effective mechanisms for adaptation of these non-linear models (EUNITE, 2000). This constitutes a still unresolved problem, with significant theoretical and practical implications: it will make a further important step in development of (non-linear) control theory and will be a theoretical basis for diverse engineering applications ranging from process modelling, control and fault detection and diagnostics to robotics and bio-informatics. Its successful solving will make possible creation of *evolving intelligence* and systems possessing it. This book presents a step in this direction.

1.3
Intelligent Adaptive Systems – a Higher Level of Control

Flexible models could be very important in cases when human is an exclusive part of the system as it is in decision making, risk assessment, Indoor Climate Control (ICC) systems, consumers behaviour modelling etc. Such systems are not *purely* technical, based on physical laws only. Practically, they incorporate subjectivity, preferences, and human behaviour.

Therefore, they need a special treatment. Applying typical (technical) system control approaches as is the current practice results in ignorance of sometimes important characteristics of such systems and eventually in poorer or far from optimal performance.

Generally speaking, similar problem exists in the so-called consumer-oriented systems (systems in which consumer's behaviour is determining the control strategy). There are emerging new types of such systems like *intelligent* agents, *intelligent* systems and various Internet applications.

A general theoretical basis for tackling this problem, however, does not exist, and different *soft computing* techniques (neural networks, fuzzy sets, and genetic algorithms) are usually used (Bigus and Bigus, 1998). The problems of models' and systems' adaptation, however, are paid little attention and are normally substituted with re-training from scratch or quasi-adaptive schemes.

Adaptive *flexible* systems able to learn *on-line* behaviour of the object of modelling and control could be named *smart adaptive systems*. They could be considered as a basis for the third level of control with the first being the local controllers, second being *conventional* adaptive controllers including supervisory controllers in hierarchical systems (Fig. 1.1).

The distinctive element is the ability **to learn, to change and enrich** their **structure *on-line***. They also could be named *intelligent* adaptive systems.

1.4
Structure of the Book

The rest of the book is presented in three parts:

➢ In the first part the basic principles of systems modelling are presented;

This introductory part sets the terminological basis for further considerations.

- It makes brief presentation in its first chapter (Chapter 2) of the basic *conventional* model types highlighting their range of effective use.
- Basic *flexible* model types are presented in the next Chapter 3.

➢ Second part is focusing attention on the problems of identification of *flexible* rule-based models. It represent the newly introduced approaches to non-linear *off-line* and quasi-linear *on-line* identification of FRB models

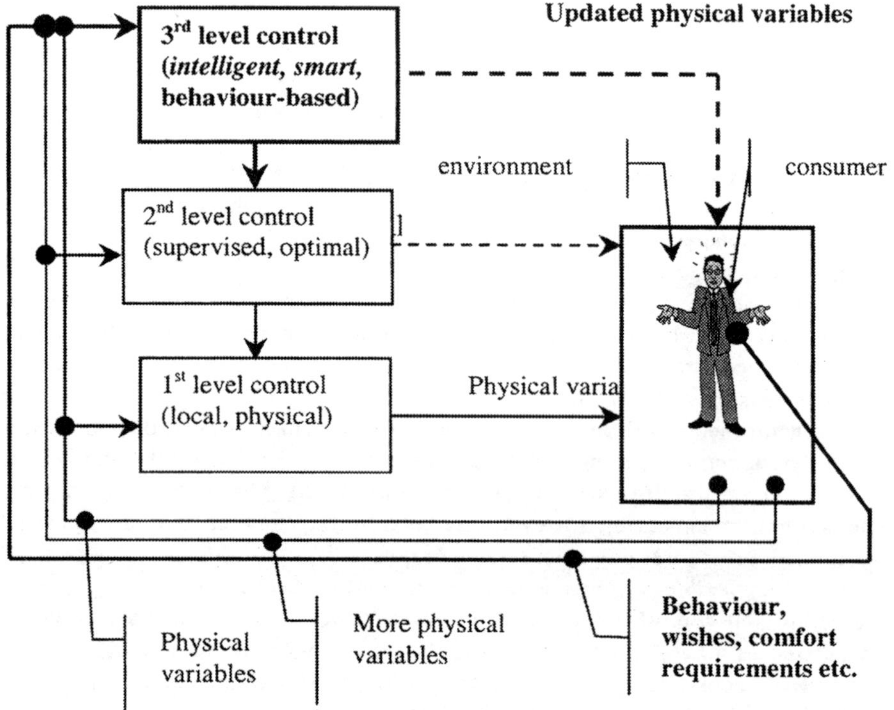

Fig. 1.1. *Intelligent* or *smart* adaptive system

- In the Chapter 4 the non-linear approach to (*off-line*) identification of *flexible* rule-based models is outlined. Two original methods for encoding *flexible* rules and linguistic terms by their indices have been represented. A new crossover operator based on the centre-of-gravity paradigm has also been introduced;
- Chapter 5 is focused on the quasi-linear approach, which makes use of the specific nature of so called Takagi-Sugeno models. Clustering and least squares techniques, which are basic in this approach are also represented briefly;
- In the Chapter 6 the notion about so called *intelligent* systems and *smart* adaptive systems has been introduced and discussed. ***E***volving **R**ule-based (***e*R**) models considered later in the book are seen as an effective tool for design of such systems;

- The newly introduced approach for *on-line* identification of Takagi-Sugeno models and ***e*R** models as an efficient tool for design and analysis of *flexible* adaptive systems are considered later on in Chapter 7:
  - ✓ One of the two basic mechanisms (rule-base *innovation*) is considered in a separate section with some illustrative examples.
  - ✓ The other basic mechanism, recursive *non-iterative* parameter up-date, is discussed later.
  - ✓ Up-date of the informative potential of new (collected in *real-time*) data is considered next;
  - ✓ *'Learning trough experience'* possible with ***e*R** models is an interesting and promising feature of this type of models;
  - ✓ A mechanism for model simplification based on similarity between the rules and linguistic terms is presented next;
  - ✓ A control algorithm based on ***e*R** models is considered and illustrated with an example of an ICC systems;
  - ✓ Basic stages and flow-chart of the algorithm are given at the end.

➢ The last, third part of the book represents some engineering applications of the proposed methodology.

It illustrates viability of the method, efficiency of the algorithms and the potential of the approach for tackling real-life problems in:

- Indoor climate control. ***e*R** models of components of ICC systems (coils, fans, boilers) has been presented based on real experimental data. Performance simulation using ***e*R** models has also been discussed and illustrated;
- *On-line* modelling of fermentation processes by ***e*R** models.
- Risk assessment. Problems of creditworthiness assessment under uncertainties of *fuzzy* type have been considered. A system for *intelligent* risk assessment in civil aviation and tendering of large-scale international construction projects has also been discussed.

These engineering examples should not be seen as limiting to the scope of possible applications of the approach.

PART I SYSTEM MODELLING: BASIC PRINCIPLES

2 CONVENTIONAL MODELS

Design of a control system and performance analysis of an object is practically impossible without having in some form a model of this object. The model could differ from the original object by structure, form etc., but it has to represent its reaction to certain input signals.

There could be, in principle, more than one model to the same process. In general, the model(s) could serve to a different purpose:

- ✓ to predict object reactions;
- ✓ to control it;
- ✓ to detect faults;
- ✓ to study the process etc.

Historically, first models have been developed in physics few centuries ago and they represent mathematically the basic, first principles of, mainly, mechanical and, later, electromagnetic behaviour of objects. During the last few decades mathematical models have been developed to such *soft* fields like biology, medicine, social sciences, economics etc.

For many practical processes, however, design of such a model is a tedious and often an expensive (time wise and computation wise) task. Therefore, a popular engineering approach is to approximate object behaviour with a non-linear curve and to fit its parameters.

There are several approaches of the second group with the latest, neural networks, becoming more and more popular. They are briefly represented in the next sections in the sake of completeness and in order to facilitate the comparison with *flexible* models, considered later.

2.1 First Principles Models

This approach supposes a thorough understanding of the nature and behaviour of the object being modelled. Physically, it is based mainly on the fundamental principles of mass- and energy conservation and their balance in closed systems. Mathematical representation of these models usually takes form of system(s) of algebraic or differential equations.

In cases, when sizes are significant, parameters of these models are distributed in space. Often, time distribution and gradients are necessary to be taken into account. From the other hand, sometimes simplifications are possible and discretised versions (difference equations) are considered.

Two engineering examples are presented here with illustrative purpose.

2.1.1 Heating/cooling Coil Model

Heating and cooling coils are commonly used component of Indoor Climate Control (ICC) systems. A general scheme of such a coil is represented in Fig.2.1

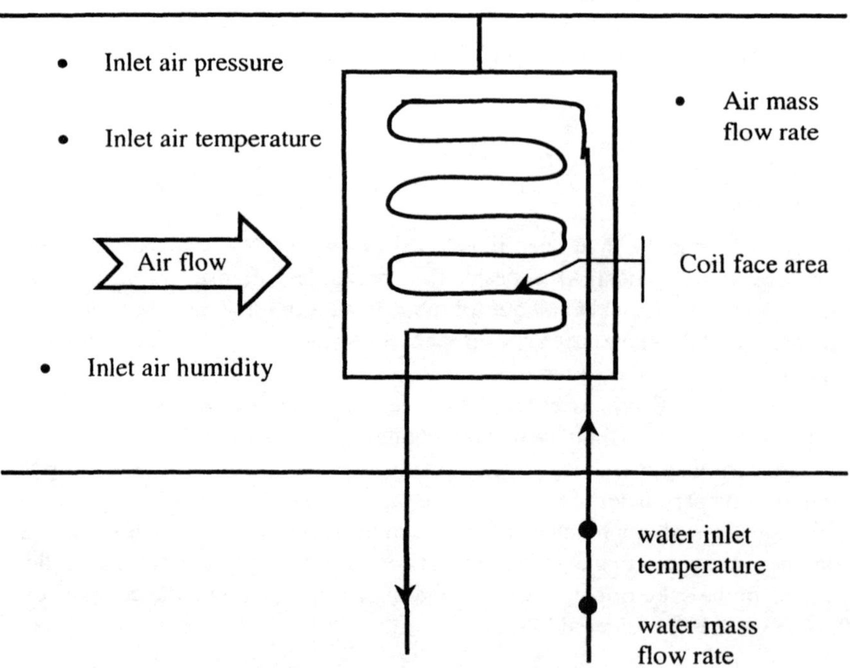

Fig.2.1. Schematic representation of a heating/cooling coil

The first principle model consists of four non-linear algebraic equations describing the mass- and energy balance in a steady state (Hanby and Wright, 1989):

$$m_a(h^{out} - h^{in}) = c_{\min} eff(T_w^{in} - T_a^{in}) \tag{2.1}$$

$$m_a(h^{out} - h^{in}) = c_w(T_w^{in} - T_a^{in}) \tag{2.2}$$

$$g^{in} - g^{out} = \frac{(1 - SHR)(T_a^{in} - T_a^{out})}{2400\ SHR} \tag{2.3}$$

$$p^{in} - p^{out} = \frac{G^2 \upsilon f}{2\ flfa} \tag{2.4}$$

where h denotes enthalpy;
p denotes air pressure;
g stands for moisture content of the air;
m_a is air mass flow rate;
T_a denotes air temperature;
T_w is water temperature;
SHR denotes sensible heat ratio;
υ denotes specific volume of the air;
$flfa$ denotes free flow area/air side area;
G denotes air mass velocity;
c_w denotes water side capacity rate;
c_{min} denotes minimum fluid capacity rate;
eff denotes coil effectiveness;
f denotes friction factor;
$_{out}$ denotes outlet;
$_{in}$ denotes intlet.

In order to determine outlet air and water temperatures, total heat transfer rate or pressure drop it is necessary to solve this system of equations numerically. Inlet air and water temperatures, air moisture content as well as air and water flow rates are usually known as outputs of a previous component of the ICC system.

Numerical solution is found usually by iterative search procedures (Newton-like gradient-based, simplex method successive substitution etc.) and is approximate, although the precision could be controlled and pre-specified. It is time-consuming, especially in performance simulation when, for example for a whole year simulation, the component model instance is called thousands of times by the main routine (Angelov et. al., 2000).

2.1.2 Fermentation Process Model

Another typical example represents a fermentation process model.

Let a general fermentation process is considered, in which a cell mass with concentration X grows in a reactor (Fig. 2.2). Starting from initial *inoculate* concentration X_0 consuming substrate, which have initial concentration S_I and is added into the reactor with dilution rate D, the cell mass grows with grow rate μ_X.

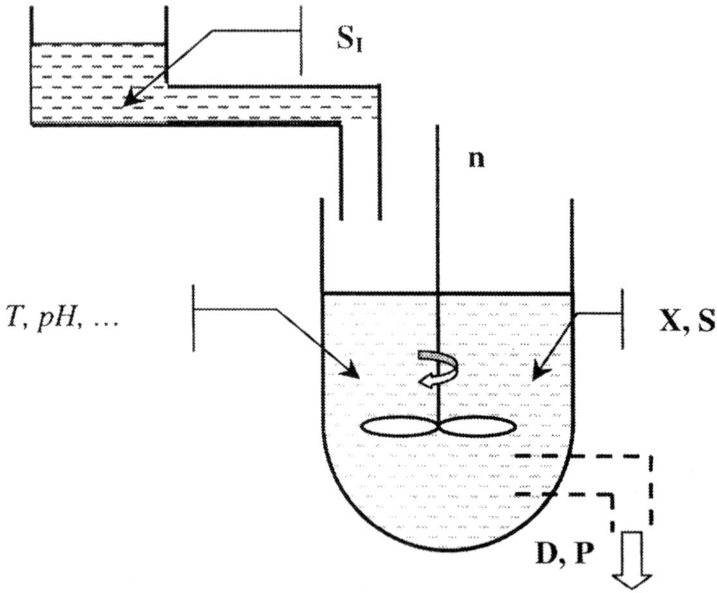

Fig 2.2. Schematic representation of a fermentation process in a stirred tank reactor

First principles models of fermentation processes are based on mass- and energy balance and are represented as sets of differential equations[1]:

$$\frac{dX}{dt} = \mu_X X - \boldsymbol{DX} \tag{2.5}$$

$$\frac{dS}{dt} = -q_S X + \boldsymbol{D(S_I\text{-}S)} \tag{2.6}$$

$$\frac{dP}{dt} = q_P X - \boldsymbol{DP} \tag{2.7}$$

where *X* denotes cell mass concentration
S denotes substratum concentration
P denotes product concentration
μ_X denotes specific growth rate
q_S denotes specific substratum consumption rate
q_P denotes specific product synthesis rate.

[1] The argument in bold is present in fed-batch and continious processes only

Very often there are other equations similar to the second equation (2.6), which represent other substrates, like oxygen, nitrogen, glucose etc. The complexity of this expression is actually hidden in the specific rates, which are, generally, highly non-linear functions of many parameters (X, S, P, temperature, pH etc.).

Practically, only a small number of parameters (usually less than four) is taken into account because of computational difficulties in parameter identification (Linko, 1988). As a result, the complicated process, which combines interactions with various natures (biological, chemical, biochemical, and physical), is not fully and adequately represented (Staniskis et.al, 1988).

This expression (2.5)-(2.7) is computationally very expensive, especially for optimisation and control and, therefore, many assumptions and simplification have been made to simplify it (Angelov et. al., 1996).

For example, in an optimal control problem based on such a model transversally (boundary) condition has to be defined. Normally it is formulated as:

"*Concentration of the feed substratum at the end of the batch fermentation* (S_{end}) *to be zero*":

$$S_{end} = 0 \tag{2.8}$$

In reality, it is an idealization, as substratum concentration never could reach this value. A more general formulation based on *fuzzy* sets has been introduced in (Angelov, 1993).

A hybrid model of this type, which uses neural networks for specific rate representation, is considered in the next section.

2.2 *Black-box* Models

Parameters of the first principles models are, however, practically time varying, non-linearly dependent on input variables and often distributed. The intrinsic nature of many processes, like decision making, cell growth and stress, human's comfort perception etc. are incompletely understood. Thus, so called *black box* approach is often used in engineering practice.

2.2.1 Linear *Black-box* Models

The simplest *black-box* model structure is the linear difference equation, which could be represented in a vector form as (Astrom and Wittenmark, 1990):

$$x(k+1) = Ax(k) + Bu(k) \tag{2.9a}$$

$$y(k) = Cx(k) + Du(k) \tag{2.9b}$$

where k denotes time instances
$y(k)$ is the output signal
$u(k)$ is the input signal
$x(k)$ denotes the state of the system
A, B, C, and D denote parameters

This linear *black-box* model is widely used in control engineering. Very often more complicated systems and processes have been *linearised* by transformation to (2.9a)-(2.9b). This is made by a proper estimation of parameters A, B, C, and D.

The problem of non-linearity and the complexity, however, normally remain hidden in the nature of the parameters, which became *time varying*. In robust systems the intervals of possible tolerance around the values of parameters are considered.

The main problem in identification of linear *black-box* models is the effective parameter estimation. Although a simplification of the reality, they play a significant role in modelling and control, because their simplicity makes possible effective *recursive* and *non-iterative* procedures like least squares to be applied in *real-time* (Strobach, 1990).

2.2.2 Polynomial Models

Polynomial models are non-linear in nature, but their structure is pretty clear and, therefore, they has found application in engineering practice for data compression (Wright, 1991), for expression of highly non-linear and non-structured dependencies for both steady state and dynamical models.

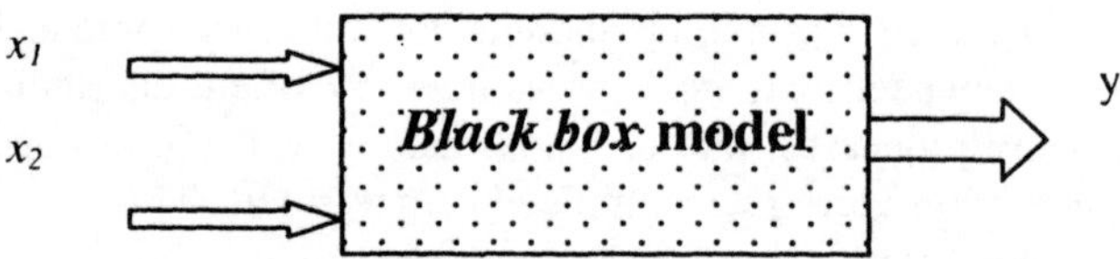

Fig. 2.3. Polynomial *black-box* model with two inputs and one output

The general form of a polynomial model with two inputs (x_1 and x_2) and an output (y) could be given by:

$$y = a_0 + a_1 x_1 + a_2 x_2 + a_3 x_1 x_2 + a_4 x_1^2 + a_5 x_2^2 + a_6 x_1^2 x_2^2 + \dots \tag{2.10}$$

It is accepted engineering practice to rely on polynomial models for calculation of efficiency of and pressure rise across centrifugal and axial fans (see the latest AHSRAE[2] toolkit, Brandemuehl et. al., 1998):

$$\psi = a_0 + a_1\phi + a_2\phi^2 + a_3\phi^3 + a_4\phi^4 \tag{2.11}$$

$$\eta_s = b_0 + b_1\phi + b_2\phi^2 + b_3\phi^3 + b_4\phi^4 \tag{2.12}$$

where ψ denotes normalised pressure;
ϕ denotes normalised airflow;
η_S denotes fan efficiency

Flow and pressure depend on the fan geometry (diameter, blade angle), speed and air density. Efficiency depends on both flow rate and pressure, density of the air and the shaft power (Brandemuehl et. al., 1998). Schematically, these models could be represented with the *black boxes* as in the Fig.2.4.

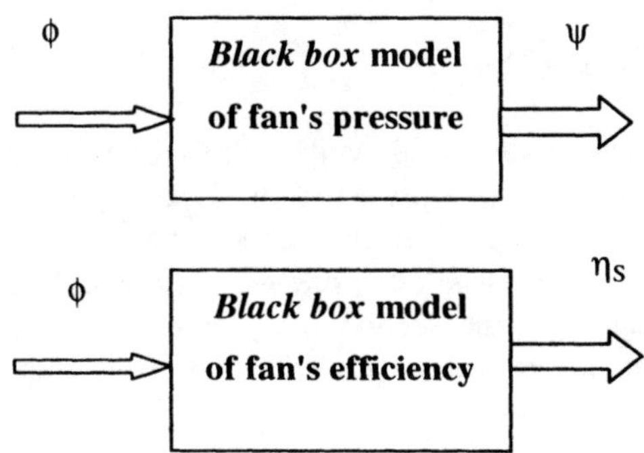

Fig. 2.4. Polynomial *black-box* models of fans pressure and efficiency

Parameters a_i and b_i $(i=0,1,\ldots 4)$ are determined based on the experimental data or are provided by the manufacturer.

2.2.3 Regression Models

Regression models could, in general, be non-linear, but without loss of generality we will consider briefly only the linear regression model called ARMAX (**A**uto-**R**egression **M**oving **A**verage with an e**X**ogenous signal). It is represented as follows (Astrom and Wittenmark, 1990):

---

[2] ASHRAE stands for **A**merican **S**ociety for **H**eating, **R**efrigerating and **A**ir-**C**onditioning **E**ngineers

$$y(k) = a_1 y(k-1) + \dots + a_n y(k-n) = \tag{2.13}$$

$$b_0 u(k-d) + \dots + b_m u(k-d-m) + e(k) + c_1 e(k-1) + \dots + c_n e(k-n)$$

ARMAX model is often generalised in a vector form as (Astrom and Wittenmark, 1990):

$$A(q)y(k) = B(q)u(k) + C(q)e(k) \tag{2.14}$$

where *A(q)*, *B(q)*, and *C(q)* represent backward shift operator and respective parameters
e(k) denotes the noise (disturbance) signal

Shift operators (backward and forward) are used for convenience of the representation. They perform time shift with as much instances, as the power of q is (Astrom and Wittenmark, 1990):

$$qf(k) = f(k+1) \tag{2.15a}$$

$$q^{-1}f(k) = f(k-1) \tag{2.15b}$$

Regression models also found wide application, especially for dynamical processes and time-series representation. One of the main problems is to determine the order (depth) of the model (how much time steps back the information is important). The other one is to estimate the parameters. There exist effective procedures for the second sub-problem, especially for linear models, while the first one is very often decided based on the experience and knowledge about the process.

2.2.4 Neural Networks

This type of *black-box* models is intensively used and developed currently, because they have *flexible, amorphous* structure and are computationally effective. It is however, not the aim of this book to investigate them. The very basic idea will be summarised only in the sake of completeness of the presentation.

From the point of view of modelling, neural networks can be seen as a layered set of interconnected *neurons*, each of which have its activation function. Normally, this is sigmoidal function of the type (Rummelhart and McClelland, 1986):

$$Output = \frac{1}{1 + e^{-\sum_{i=1}^{n} weight_i \, Input_i}} \tag{2.16}$$

The unknown parameters (weights) are determined normally by error *back-propagation* (Werbos, 1990), which is a gradient-based algorithm or, rarely, by genetic algorithms.

2.2.4.1
Radial-basis Functions (RBF) Neural Networks

A special case of neural networks, so called RBF-NN, deserve more attention in relation to the approach considered later in the book. The reason is that they are close by definition to *flexible* rule-based models and in particular to so-called Takagi-Sugeno models. They, however, are still representative of the class of *black-box* models, including all negative characteristics, like the lack of interpretability, limited range of validity and assumption of the initial structure.

The general structure of such a network is represented in the Fig. 2.5. The input signals ($Input_1$, $Input_2$, and $Input_3$) are supplied to the hidden layer's neurones, which perform radial-basis function transformation. This is, in fact, the Gaussian function:

$$RBF\ (Input\ ) = e^{-\frac{\| Input\ - Centre\ \|^2}{2\, spread\ ^2}} \tag{2.17}$$

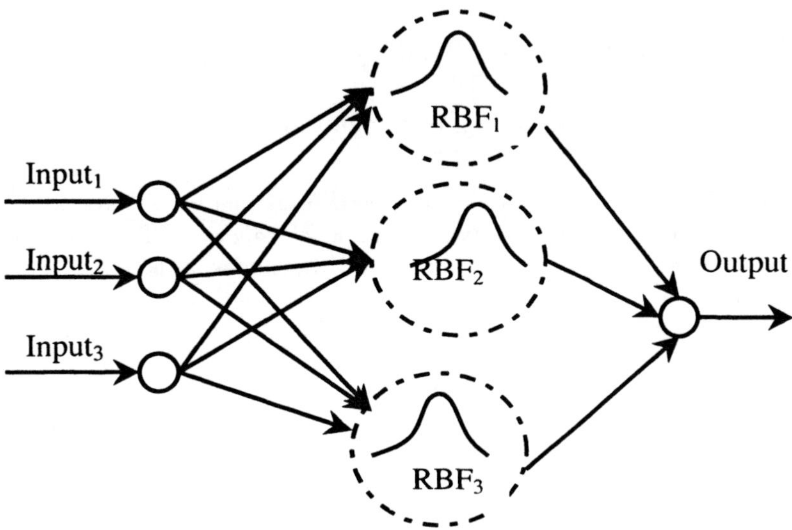

Fig.2.5. RBF neural network

The output from the hidden neurones depends on the closeness measured by the Euclidean distance between the current measured point (*Input*) and the *Centre* of the RBF. If the *Input* value is close to the *Centres* then the hidden RBF neurones are activated more significantly and the *Output* from the hidden neurones has larger value. And vice versa, if the *Input* is far away in the state space from the *Centre* of the RBF function, then the output signal from the hidden neurones has smaller value.

The total *Output* from the network is a weighted sum of the individual RBF activations:

$$Output = \sum_{i=1}^{hidden} weight_i RBF_i \tag{2.18}$$

2.2.4.2
Hybrid NN - First Principles Model of a Fermentation Process

Neural networks have been used for description of the highly non-linear specific rates of the model (2.5)-(2.7) of a biotechnological process (Chen et. al., 2000):

$$\mu_X \big|_{k+1} = f_X(X_k, S_k, P_k, pH_k) \qquad k=0,\ldots,K-1 \tag{2.19}$$

$$q_S \big|_{k+1} = f_S(X_K, S_K, P_K, pH_k) \tag{2.20}$$

$$q_P \big|_{k+1} = f_P(X_K, S_K, P_K, pH_k) \tag{2.21}$$

where K denotes number of steps of discretisation,
f_X, f_S, f_P are non-linear function.

Such a neural network is trained on experimental data from a process of vanillin production from vanillic acid by *Pycnoporus cinnabarinus* (Bernard et. al., 1999) using a three-layered neural network with five neurons in the middle layer, given in the Fig. 2.6.

More details about neural networks could be found elsewhere (Rumelhart and McClelland, 1986; Werbos, 1990) and, especially, in relation with *flexible* rule-based models in (Jang et. al., 1997).

2.3
Conclusion

A brief introduction to *conventional* models has been given in this chapter. First principle and *black-box* models have been considered. More details could be found in (Ljung, 1987; Astrom and Wittenmark, 1990).

First principle models have been characterised and illustrated with two typical examples:

- Highly non-linear model of a heating/cooling coil – a typical component of any air-conditioning system;
- Complicated and highly non-linear model of secondary metabolite fermentation in a stirred tank reactor.

This type of models are preferable when the process is well known and understood. They, however, are very often computationally heavy and even prohibitive. Nevertheless, they have their field of application, in particular, in hybrid schemes in combination with other type of models (neural networks, *flexible* models etc.).

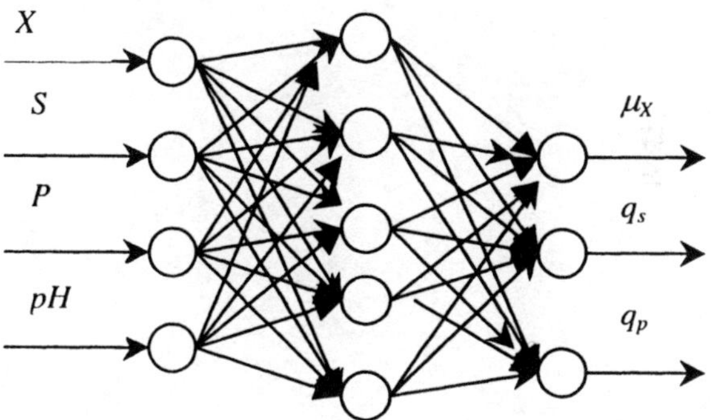

Fig.2.6. Schematic representation of a neural network for specific rates of a fermentation process modelling

Black-box models have been presented briefly as another type of *conventional* models. Linear, polynomial and regression models have been considered. *Black-box* models have been illustrated with an engineering example of fan power and efficiency modelling.

They are practically used when the process is highly non-linear, time varying and incompletely understood. They require enough (as quantity and quality) experimental data and have a number of restrictions and limitations the most important being the *lack of transparency*, limited range of validity and the necessity to know the structure of the model *a priori*.

Finally, the most recent representative of *black-box* models, so-called neural networks (NN) have been presented. Radial-basis function NN has been paid more attention since they have parallels with *flexible* rule-based, and especially with so-called Takagi-Sugeno models.

An example of their application to modelling specific rates of a fermentation process has been given with illustrative purposes.

3 FLEXIBLE MODELS

First principle-based and *black-box* models have some limitations, which have been mentioned. In a pursuit to overcome it an alternative, which have ***flexible*** enough structure to represent adequately non-linearity and uncertainty of real processes **and** is **transparent** enough to be easy for inspection, analysis, incorporation of existing knowledge and suppression of undesired one, represent *fuzzy* models.

They have been developed during the last two decades as a result of an interaction of the *Fuzzy* Set Theory and the Control Theory.

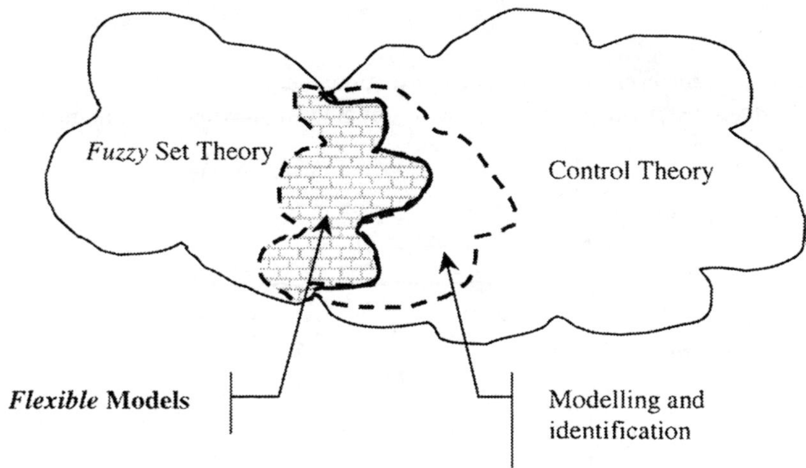

Fig. 3.1. *Flexible* Models as an intersection of two basic theories

Some definitions, forming the terminological basis and ideas, which will be used later, are briefly presented in this chapter.

3.1 *Fuzzy* Set Theory - Basic Introduction

The key notion of *fuzzy* set theory is the **membership to a set**. In classical set theory, foundations of which have been set up long ago, each element x of universe U has two options only in respect to a certain set S:

- it could **belong** ($\in$) to S;
- it could **not belong** ($\notin$) to S.

This principle, called the *Law of excluded middle* has been formulated by Aristotle.

In *fuzzy* set theory (Zadeh, 1965) this principle has been revised such that an element x belongs to the *fuzzy* set F with certain degree of membership μ_F (it is accepted that this degree is a real number, laying in the unity interval):

$$\mu_F \in [0;1] \tag{3.1}$$

Example 3.1

The set *Positive temperature* (in °C) is a conventional, crisp, non-fuzzy set;

Set *Medium temperature* is a *fuzzy* set (Fig. 3.2), having certain membership function μ_{MT}.

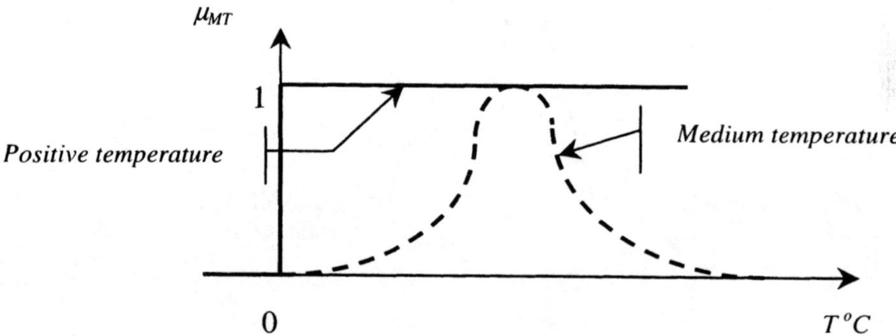

Fig. 3.2. *Fuzzy* and *conventional* sets defining temperature

It should be noted, that while a *fuzzy* set suppose non crisp, non sharp boundaries of membership, it is in the same time very well defined by the membership function and is not *fuzzy* in the literal sense of this word. After the membership function is defined, it is a well-shaped and well-determined set, although it allows partial and combined membership. This is the reason the author prefers and will use the term *flexible* as a synonim to *fuzzy* in this sense.

3.1.1 *Fuzzy* Set Definition

Definition 3.1

A *fuzzy* set F is defined in U by:

$$F = \{(x, \mu_F(x)) | x \in U\} \tag{3.2}$$

where $\mu_F: U \to [0;1]$ is called the *membership function* of F
$\mu_F(x)$ is the degree of membership, with which x belongs to F
U is the inverse of discourse.

Each *fuzzy* set is uniquely determined by its membership function. Different mathematical functions are used in practice to represent the membership function:

- Gaussian;
- piece-wise linear;
- triangular;
- trapezoidal;
- sigmoid etc.

Two of the most often used types are triangular and Gaussian. Gaussian curve is given by:

$$\mu(x) = e^{-\frac{(x-c)^2}{2\sigma^2}} \tag{3.3}$$

Membership function itself is defined by its parameters. For example, the Gaussian-type membership function ($\mu(x)$) is defined by its centre (c) and spread (σ) as shown in the Fig.3.3.

Definition and parameter identification of membership functions constitute the trickiest issue in the practical use of *fuzzy* sets theory. There are a lot of discussions and no common fundamental recipe for doing this. Data-driven techniques and their combination with expert knowledge represent one possible solution, which in the point of view of the author is most promising.

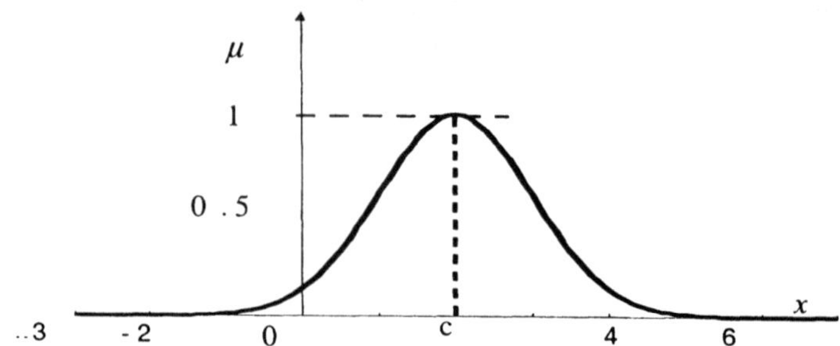

Fig. 3.3. Gaussian type membership function

3.1.2 Basic Operations over *Fuzzy* Sets

Without going into details the basic operations over *fuzzy* sets will be presented briefly as a basis for further considerations.

3.1.2.1 T-norms

One of the basic operations over *fuzzy* sets is the conjunction of two *fuzzy* sets. It is equivalent to the logical AND operation (Klir and Folger, 1988):

$$\mu_A \bigcap \mu_B = A \textit{ AND } B \tag{3.4}$$

The most commonly used mathematical representation is proposed in the pioneering paper on *fuzzy* sets (Zadeh, 1965):

$$\mu_A \bigcap \mu_B = \min(\mu_A, \mu_B) \tag{3.5}$$

Another definition, which is used very often, is the so-called *product operator*:

$$\mu_A \bigcap \mu_B = \mu_A * \mu_B \tag{3.6}$$

3.1.2.2
S-norms

The next basic operation over *fuzzy* sets is the union. It is equivalent to the logical OR operation (Klir and Folger, 1988):

$$\mu_A \bigcup \mu_B = A \textbf{ OR } B \tag{3.7}$$

Again, the most commonly used mathematical representation have been proposed in the first paper on *fuzzy* sets (Zadeh, 1965):

$$\mu_A \bigcup \mu_B = \max(\mu_A, \mu_B) \tag{3.8}$$

Another definition, which is used very often, is the so-called *sum operator*:

$$\mu_A \bigcup \mu_B = \mu_A + \mu_B - \mu_A \mu_B \tag{3.9}$$

3.1.2.3
Negation

The last of the basic operations over fuzzy sets is the complement of a *fuzzy* set. It is equivalent to the logical NOT operation (Klir and Folger, 1988):

$$\neg \mu_A = NOT(A) \tag{3.10}$$

There is only one mathematical representation of this operator:

$$\neg \mu_A = 1 - \mu_A \tag{3.11}$$

There are several functions, which also are considered as T- and S-norms, including parameterised ones. All of them have to posses some properties, like to be (Klir and Folger, 1988):

- Monotonic;
- Commutative;
- Associative;
- Distributive;
- Idempotent etc.

Practical use, however, found the basic operations (3.4)-(3.11).

3.1.2.4 De-fuzzification

De-fuzzification is another basic operator, which performs mapping of a *fuzzy* set into a crisp number. This operator is essential in *flexible* logic controllers and in all engineering systems, when the output have to be a crisp, non-*fuzzy* number. It could be seen as analogy to the mathematical expectation in probabilistic sets.

The two mostly used operators are so called *centre-of-area* (CoA) and the mean of maximums (MoM). *CoA* operator returns a point, which lays on the line dividing the area below the membership functions into two equal parts:

$$x^* = \frac{\sum_{i=1}^{card(x)} \mu(x_i) x_i}{\sum_{i=1}^{card(x)} \mu(x_i)} \tag{3.12}$$

where *card(x)* denotes the cardinality of *x*

MoM operator returns the mean of all maximum points of the membership function (if there is more than one maximum):

$$x^* = \frac{\sum_{i=1}^{m} \arg \max_{i=1}^{card(x)}(\mu(x_i))}{m} \tag{3.13}$$

where m denotes the number of maximums

They have been generalised with a parametric operator called BADD (from BAsic De-fuzzification Distribution) introduced by Filev and Yager (1991). It includes CoA (when parameter $\theta = 1$) and MoM (when parameter θ tends to infinity) as special cases.

Parameter θ is used as additional leverage for tuning. It expresses the balance and proportion of contribution of each particular *flexible* rule to the final output of the model.

$$x^* = \frac{\sum_{i=1}^{card(x)} \mu^{\theta}(x_i) x_i}{\sum_{i=1}^{card(x)} \mu^{\theta}(x_i)} \tag{3.14}$$

This operator will be used further as a more general one (for $\theta = 1$ it supposes CoA operator).

3.1.2.5
Degree of Similarity Between Fuzzy Sets

The degree of similarity between two fuzzy sets *A(x)* and *B(x)* is usually defined as a proportion of the intersection of *A(x)* and *B(x)* (*A* **AND** *B*) to the union of *A(x)* and *B(x)* (*A* **OR** *B*) over all possible values of the independent variable (*x*), i.e. cardinality of A and B (Klir and Folger, 1988):

$$Sim_{A,B} = \frac{|A(x) \bigcap B(x)|}{|A(x) \bigcup B(x)|} \tag{3.15}$$

The following two special cases are quite obvious:

- Equal *fuzzy* sets ($A(x) = B(x)|_{card(x)}$) yields $Sim_{A,B} = 1$;
- Non-overlapping *fuzzy* sets yields $Sim_{A,B} = 0$.

3.2
Models with *Flexible* Parameters or (In)equalities

This type of models has been used for description of *flexible* constraints mainly in optimisation problems, but is applicable in process modelling and control in general (Angelov and Tzonkov, 1993). It treats parameters and (in)equality relations in *conventional* models as *flexible* sets.

3.2.1
Models with *Flexible* Parameters

The logic behind this type of *flexible* models is that value of a parameter of a model is uncertain and that it could be represented by a *flexible* set.

For example, if we consider the regression model of the pressure across an axial or centrifugal fan (2.11)-(2.12), we could suppose that value of parameters are represented by *flexible* sets with their respective membership functions

$$\psi = \tilde{a}_0 + \tilde{a}_1 \phi + \tilde{a}_2 \phi^2 + \tilde{a}_3 \phi^3 + \tilde{a}_4 \phi^4 \tag{3.16}$$

where $\tilde{a}_i$ $i=0,1,2,3,4$ are *flexible* numbers.

Normally, *flexible* parameters are represented by triangular membership functions (Orlovski, 1978), Fig.3.4:

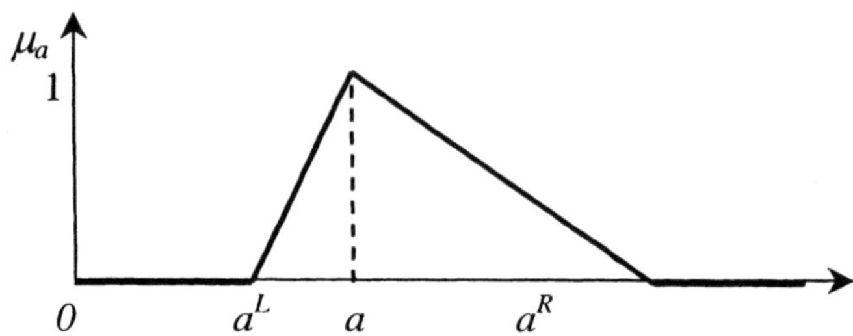

Fig. 3.4. *Flexible* parameter $\tilde{a}_i$

where a^L denotes left boundary of the *flexible* parameter a;
a^R is its right boundary;

This type of models is used predominantly in optimisation problems (Tanaka and Asai, 1984; Rommelfanger, 1989; Carlsson and Fuller, 2001) and will not be considered further in this book.

3.2.2 Models with *Flexible* (In)equalities

This type of *flexible* models is used mainly to represent constraints in optimisation problems. It has been introduced in (Zimmermann, 1983) for representation of *flexible* objectives and constraints in a *flexible* linear programming problem. Filev and Angelov (1992) introduced them for the optimal control problem considering *flexible* transversally condition (Angelov, 1993). This type of models could be illustrated with the following

Example 3.2

The purpose of an ICC system is to ensure quality of environment of the occupants of a building zone. The *conventional* way to express this inherently *flexible* objective is to formulate a standard in crisp terms for basic physical variables (temperature and humidity) or in the terms of so called PPD or PMV[1].

This is, in fact, the current engineering practice in this field under the ISO 7730 and ASHARE-55 and ASHARE-62 standards (Taylor, 1995), which are recently an object of intensive re-consideration (Brager and de Dear, 1998). An obvious improvement is to consider *relaxed*, *flexible* comfort requirements (Kuntze and Bernard, 1998; Angelov, 1999). Then the requirement for *comfort* temperature could be personalised and could be in the form as presented in the Fig. 3.5.

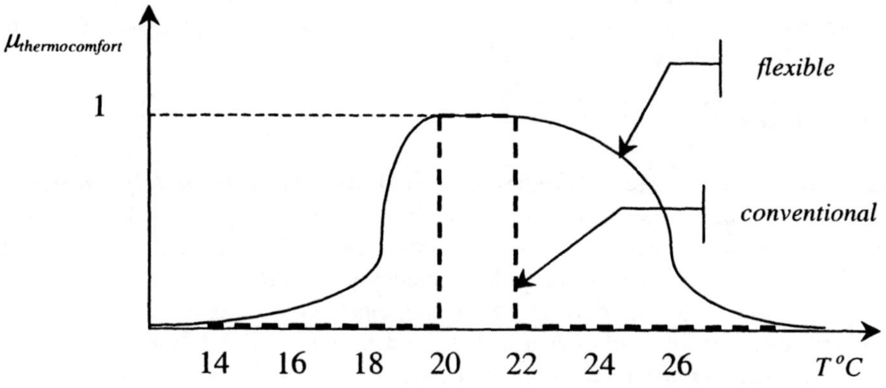

Fig. 3.5. Personalised thermal comfort membership function

For this particular case, the occupier prefers *slightly hot* environment (22-24°C) than a *slightly cool* one (18-20°C). Generally, *flexible* inequalities allow slight violation of such *soft* constraint, although the degree of acceptance decreases with the increase of the value of the violation.

Let us consider another simple

Example 3.3

The following *flexible* inequality constraint:

$$T \mathrel{\tilde{\leq}} 25^{o}C \qquad (3.17a)$$

---

[1] From **P**redicted **P**ercentage of **D**issatisfied people and **P**redicted **M**ean **V**ote

could be represented by a *fuzzy* set, which could have the following membership function:

$$\mu = \begin{cases} 1 & \ldots \quad ; T < 25\,^{o}C \\ T - 25 & \ldots ; 25\,^{o}C \le T \le 26\,^{o}C \\ 0 & \ldots \quad T > 26\,^{o}C \end{cases} \tag{3.17b}$$

where $\tilde{\le}$ denotes *flexible* inequality

Similarly, *flexible* objective functions in optimisation problems have been introduced (Zimmerman, 1983). In certain processes (like biotechnology) so-called *typical* process behaviour could be characterised. In the same time, it is not easy to quantify the influence to the product quality and process duration of factors, which define this *typical* process behaviour.

Example 3.4

In enzyme glucose-oxidase synthesis, the **culture colour** is ***slightly*** **brown**, when fermentation is *good* and it is ***yellow***, when the cell mass *has grown without significant increase of the enzyme activity* (Angelov, 1993). Similarly, the **smell** of the culture and **hyphen structure** is *specific* when the fermentation is *bad*.

In practice, the experienced (bio)technologist could easily judge whether the process behaviour is *good* or *bad* and could determine which fermentation is ***typical***, out of a large number of experimental runs. Based on this *typical* process run, the following *flexible* model description have been introduced (Angelov, 1993):

$$x_{k+1} \tilde{=} f(x_k, u_k) \tag{3.18}$$

where $\tilde{=}$ denotes *flexible* equality

This type of *flexible* models have been used basically for dynamic optimisation and optimal control and will not be consider further in this book. For more details, see (Angelov and Tzonkov, 1993).

3.3 *Flexible* Rule-based Models

For the first time *flexible* rule-based (FRB) models have been reported in (Zadeh, 1973). Generally speaking, a FRB model is a non-linear mapping of inputs to the output represented in a linguistic form:

***IF** (antecedent) **THEN** (consequence)*

Three basic groups of FRB models differ by specific elements of this representation: *flexible* or crisp inputs and output, relations between the antecedent and consequent parts. They are briefly represented in this section.

3.3.1 *Flexible* Relational Models

Flexible relational models have been introduced by Pedrycz (1984) as combinations of *flexible* relations:

$$R_i : (X \times Y) \rightarrow [0;1] \qquad i=1,2,\ldots,NR \tag{3.19}$$

where R is a *flexible* relation, mapping
NR is the number of rules

Flexible implications (*flexible* logic) are used for simulation based on this model (Driankov et. al., 1993). Alternatively, the relation is treated as a conjunction (Mamdani's method). The last approach has advantage that it allows an inversion of the model. It considers logical AND aggregation of both antecedent and consequent parts and is calculated as

$$\mu_{R_i}(x,y) = \mu_{X_i}(x) \bigcap \mu_{Y_i}(y) \tag{3.20}$$

An obvious computational inconvenience is that the conjunction (normally *min* operator) is calculated for the whole cardinality of X and Y, i.e. for all possible (x,y) pairs. The overall *flexible* relation R, which represents the whole model (combined effect of all rules)

$$R_i : IF\ (x.isX_i) THEN\ (y.isY_i) \quad i=1,2,..,NR \tag{3.21}$$

is represented by union (disjunction) of all individual relations R_i:

$$R = \bigcup_{i=1}^{RN} R_i \tag{3.22}$$

Its degree of fulfilment is calculated by applying some disjunction operator over the conjunction of particular rules:

$$\mu_R(x,y) = \max_{i=1}^{RN} \left(\mu_{A_i}(x) \bigcap \mu_{B_i}(x)\right) \tag{3.23}$$

The output of the *flexible* relational model is calculated by so-called relational *max-min* composition, denoted as °:

$$y = x \circ R \tag{3.24}$$

Details about this type of *flexible* models could be found in (Pedrycz, 1993).

3.3.2 *Mamdani*-type Models

In fact, this type of models has been originally introduced by Zadeh (1973), but it is called Mamdani type as it has been further developed by E. Mamdani (1977). Generally, it is represented in the same form (3.21), where input and output variables are linguistic (*Small, High, Almost Zero* etc.).

Since input is normally a vector, then (3.21) could be re-written as:

$$R_i : IF\ (x_1 isX_1)\ AND\ (x_2 isX_2)\ AND\ \ldots\ AND\ (x_n isX_n)$$

$$THEN\ \ (y.isY);\ldots\ i = 1,2,\ldots,\ NR \tag{3.25}$$

where x_i denotes a *flexble* linguistic input variable;
y stands for the output *flexible* linguistic variable;
Y denotes the linguistic term of y; $Y \in \{Y^1;Y^2;\ldots;Y^{m_0}\}$;
X_i; $X_i \in \{X_i^1; X_i^2;\ldots; X_i^{m_0}\}$; denotes the linguistic term of the ith input variable

Each linguistic term is defined by a *flexible* set and a membership function, which itself is determined by its parameters (these could be centres and spreads of Gaussian or bell-type functions, co-ordinates of the apexes of triangular functions and singletons). The rule base $R = \{R_i | i = 1,2,\ldots,\ NR\}$ forms together with the *flexible* sets the knowledge base of the linguistic model.

The number of all possible *flexible* rules (complete set) for a specified number of linguistic variables and their linguistic terms is extremely high for realistic dimensions (some tens of linguistic variables and linguistic terms), because of the combinatorial explosion called *curse of dimensionality* (Yager and Filev, 1994). The following expression defines the number of all possible rules without taking into account the sub-sets of rules:

$$TNR = \prod_{j=1}^{n+1} m_j \tag{3.26}$$

where m_j is the number of linguistic terms of the jth linguistic variable;
TNR - total number of all possible full rules

For example, the number of all possible rules which could be formed using 6 variables (*n=6*) with 9 linguistic terms each is more than *4* millions (*TNR*=9^7=*4,782,969)!* If take into account rules with so called 'don't care element' or 'wild card', when not all variables participate in each rule this number is even higher. It is practically impossible to interpret and inefficiently to use such a model, even if suppose that it is generated somehow.

In practice, the number of used rules is significantly smaller then the number of all possible rules (*NR* << *TNR*), because of information redundancy. The main driving factor in determining rules number is the precision of the final model, but the model complexity, transparency and interpretability become more and more an issue which is taken into account (Setnes and Roubos, 1999; Angelov and Buswell, 2001b).

The algorithm for calculating the outputs of the linguistic model based on certain inputs is called *fuzzy inference mechanism*. It is based on the *max-min* composition and *fuzzy* relational calculus (Pedrycz, 1993).

3.3.3 *Takagi-Sugeno*-type Models

TSK model has quite similar form to the linguistic model (3.25):

$$R_i : IF\ (x_1 isX_{i1})\ AND\ (x_2 isX_{i2})\ AND\ ...\ AND\ (x_n isX_{in})$$
$$THEN\ (y_i = a_{i1}x_1 + a_{i2}x_2 + ... + a_{in}x_n + b_i);...\ i = 1,2,...,\ NR \tag{3.27}$$

where R_i denotes the ith *flexible* rule
NR is the number of *flexible* rules
x is the input vector; $x = [x_1,x_2,\dots,x_n]^T$
X_{ij} denotes the linguistic term for the ith *flexible* rule (j=1,2,...,n);
y_i is the output of the ith *flexible* rule;
a_{ij} and b_i are parameters of the consequence

The main distinction being the form of the outputs, which in TSK model are (linear) functions of input variables. In principle, non-linear functions are also possible, but more often are used linear functions or even scalar (constant) outputs called singletons:

$$R_i : IF\ (x_1 is X_{i1})\ AND\ (x_2 is X_{i2})\ AND\ ...\ AND\ (x_n is X_{in})$$
$$THEN\ (y_i = s_i);...\ i = 1,2,...,\ NR \tag{3.28}$$

From (3.28) and (3.25) it is seen that Mamdani type of FRB models could be presented as an extension of TSK models having singleton outputs (zero order TSK models) for the case when these outputs are considered as *flexible* sets.

TSK models are often called also quasi-linear models (Filev, 1991) as, in fact, they perform a weighted combination of distinct, although overlapping linear sub-models, each of them operating in a strongly defined sub-space of states, determined by the antecedent part of the model (Fig. 3.6).

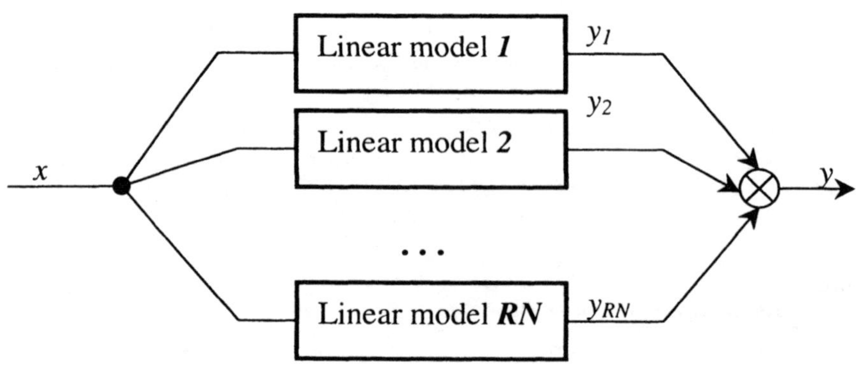

Fig 3.6. TSK as a *quasi-linear* model

Example 3.5

This is a typical one. The growth curve of cell mass in a batch or fed-batch fermentation process, represented by its concentration (Fig.3.7). For most of the processes three distinct phases exist, which represent different physiological state and the age of the culture, namely so-called “lag-phase”, “exponential growth”, and “saturation phase”.

TSK model gives a quasi-linear approximation of this process by three switching lines. The smooth transition is controlled by the membership functions as represented in the Fig. 3.8.

The model output is calculated by weighted averaging of individual rules' contribution (y_i) with the weights (ω_i) being non-linear functions of inputs, aggregating the influence of membership functions of inputs and their partial importance:

$$y = \frac{\sum_{i=1}^{RN} \mu_i(x) y_i}{\sum_{i=1}^{RN} \mu_i(x)} = \sum_{i=1}^{RN} \omega_i(x) y_i \tag{3.29}$$

where $\omega_i(x) = \dfrac{\mu_i(x)}{\sum_{i=1}^{RN} \mu_i(x)}$ is the weight of ith rule

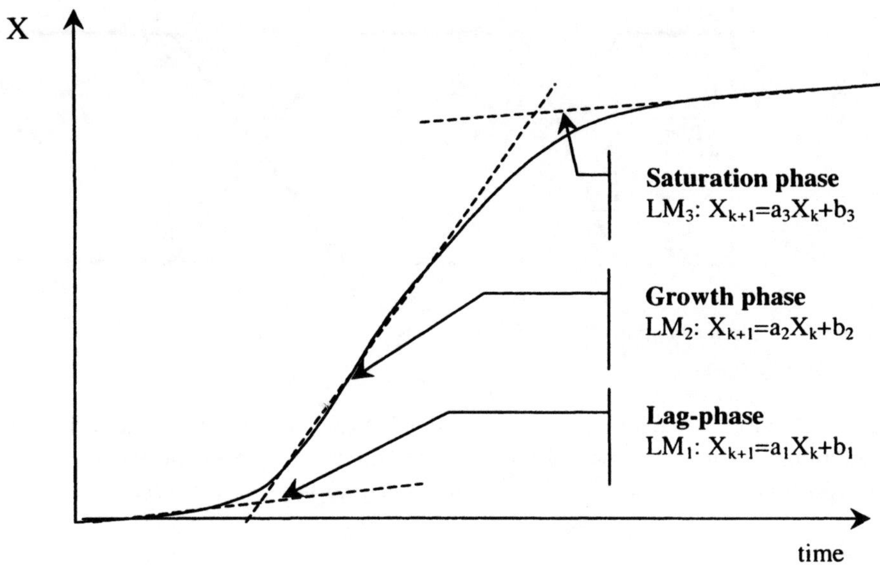

Fig. 3.7. Typical growth of a cell mass in a (fed-) batch fermentation process

For linear output sub-models we have:

$$y = \sum_{i=1}^{R} \left(\sum_{j=1}^{n} \varpi_i(x) a_{ij} x_j + \varpi_i(x) b_i \right)$$

This form of presentation of TSK models (3.29) highlights their relation to general function approximators called basis functions expansion (Friedman, 1991), to which belong also neural networks, and particularly radial basis functions, splines etc. TSK model with singletons as output functions ($y_i = b_i = constant$) also belongs to this class of functions (Jang et. al, 1997).

The weighted averaging (3.29) is similar to CoA de-fuzzification operation over *flexible* set (3.12). The difference is that it is performed over rules, which have cardinality equal to their number (*NR*).

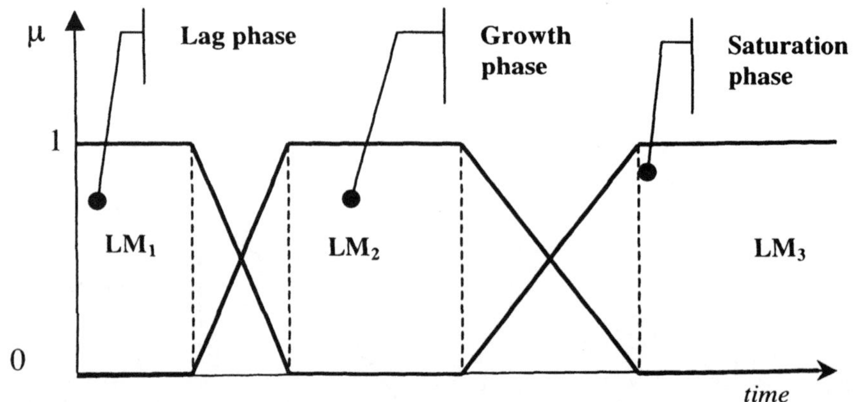

Fig.3.8. Membership functions of TSK model

A parameterised weighting, similar to BADD (3.14) is used here as a leverage for additional *flexibility*, which allows varying the proportion of the contribution of the rules with the higher degree of fulfilment in respect to all other rules:

$$\omega_i(x) = \frac{\sum_{i=1}^{RN} \mu_i^{\theta}(x)\, x_i}{\sum_{i=1}^{RN} \mu_i^{\theta}(x)} \tag{3.30}$$

In fact, this *soft* de-fuzzification operator makes TSK models close to relational models in the sense that the output is not restricted to the inputs grid partition only, but depends on additional parameter (θ) also, which could be used for fine-tuning.

The degree of fulfilment of each rule μ_i is determined by applying basic operators over *fuzzy* sets of the antecedent part of the rule. For example, the degree of fulfilment of the *flexible* rule (3.31) is determined as aggregation of conjunction, negation, and disjunction, as represented in (3.32).

$$IF\ (T_a isHigh\ )AND\ (m_a isNOT\ .Small\ )OR\ (T_w isLow\ )$$
$$THEN\ \ (T_{\sup pl} isLow\ ) \tag{3.31}$$

where the output variable T_{suppl} denotes the supply air temperature.

$$DoF\ = \mu_{T_a}(1 - \mu_{ma})\mu_{T_w} \tag{3.32}$$

where *DoF* denotes the Degree of Fulfilment of the respective *flexible* rule

TSK models could be used also to model the dynamical behaviour of an object as a *flexible* version of the non-linear auto regressive model:

$$R_i\ :IF\ (y_k isY_{i1})AND\ (y_{k-1} isY_{i2})AND\ ...AND\ (y_{k-n+1} isY_{in}) \tag{3.33}$$

$$AND\ (x_k isX_{i1})AND\ (x_{k-1} isX_{i2})AND\ ...AND\ (x_{k-m+1} isX_{in})THEN\ \ (y_{k+1} isY_i)$$

In this way, discretised state-space dynamical model (2.9a)-(2.9b) could be represented specifying the current state of the system (x_k) and the input (u_k) as inputs to this model determining the next state (x_{k+1}) as an output.

3.4 Conclusion

In this chapter a basic introduction to the *fuzzy* sets is given as well as the basic operations over these sets have been defined. Some of them, like logical AND, de-fuzzification, and similarity measures play an important role in the approach presented in the second part of the book. An illustrative example of a definition of temperature by a conventional and by a *fuzzy* set is also given.

In the next section *flexible* models have been represented as an alternative to the conventional ones. Models with *flexible* inequalities and/or equalities as well as models with *flexible* parameters have been briefly presented. They are used mainly in decision making and optimisation and are not considered further in the book.

Basic types of *flexible* rule-based models, in which *fuzzy* sets have been used to represent relations, inputs or the output, have been outlined. Models with *flexible* relations are presented briefly, as they also are not considered later in the book.

So-called Mamdani models are very close to Takagi-Sugeno (TSK) models in nature and differ by the form of the output only. They could be used in the so-called ***non-linear*** approach to identification considered in the next chapter, but are presented in less detail than TSK models as they are not appropriate for ***recursive*** *on-line* identification, considered in the Chapter 7. Illustrative examples from indoor climate control systems and biotechnology are given.

More attention is given to the so-called Takagi-Sugeno models, as they will be further considered in the book, because of their computational efficiency and convenience. The dualistic nature of TSK models being both non-linear as a whole and locally linear by sub-models has been represented and illustrated by an example from biotechnology in most general form. This quasi-linear character of TSK models is used in the approach for *on-line* identification and rule-base evolution, presented in the Chapter 7.

Parameterised de-fuzzification (BADD) operator is used to form the output of TSK models, which includes the centre-of-area one as a special case. In this way an additional parametric leverage is defined, which allows for finer tuning and optimisation.

For more details on *flexible* models, please refer to (Driankov et. al., 1993; Yager and Filev, 1994).

PART II *FLEXIBLE* MODELS IDENTIFICATION

The problem of model identification is traditionally related to the problem of control, although models could be used for other purposes as well. For example, they could be used for classification, decision support, fault detection and diagnostics etc.

The identification basically includes the following four phases (Astrom and Wittenmark, 1990):

- A. Experimental planing
- B. Model structure selection
- C. Parameter estimation
- D. Validation

The most important are phases B and C, while the phase D proves the viability of the resulting model and has high practical importance. The structure selection (phase B) is often based on prior knowledge and is to some extent a subjective process.

Highly desirable is an algorithm to combine both model structure selection (phase B) and parameter estimation (phase C). Normally, identification is an *iterative* procedure, although (especially for linear models) *non-iterative* approaches for identification also exist (Chiu, 1994).

The identification of *flexible* models has its beginning with the corner-stone paper by Takagi and Sugeno (1985) in which the *quasi-linear* TSK models have been introduced. This problem has been treated later in (Xu and Lu, 1989; Filev, 1991; Sugeno and Yasukawa, 1993 etc.).

The problem of identification differs when the purpose of the use of the model is classification and when it is control. Computational efforts, dynamics, and stability are the main issues, which matter in control, while the lower precision of linear models could be compensated to some extent with the control schemes operating normally in *real-time* and *on-line* mode.

The adaptation is of higher importance for control-oriented models than precision itself. This could be seen as a reason why linear PI and PID controllers and linear (often regression) models are still widely used in practice.

A thorough and detail study on (*off-line*) identification of *flexible* models specifically in relation to control has been given in (Yager and Filev, 1994). They introduced also the pivotal clustering method called *mountain clustering* (Yager and Filev, 1993), modified later in (Chiu, 1994). This clustering method surpasses others, like fuzzy C-means, with its simplicity and efficiency and have been used later as a basis of a fast and robust algorithm for *flexible* models identification in combination with the linear least squares.

The problem of identification of *flexible* models, consists of the same basic phases. Two of them are most important:

- Structure identification.

In the case of *flexible* models, it means to identify

- ✓ input and output variables;

- ✓ number of rules and linguistic terms;
- ✓ type of membership functions

➢ Parameter identification.

This includes basically the following sub-problems:

- ✓ Identification of parameters of membership functions of the antecedent part of the rules;
- ✓ Identification of parameters of the consequent (linear or singletons) part of the rules;
- ✓ Identification of the parameter of the generalised BADD de-fuzzification method, if used.

In general, there are two approaches to identification of *flexible* models:

✓ ***non-linear*** approach

This approach is based on the fact, that *flexible* rule-based models are non-linear input-output mappings. Therefore, it relies on non-linear optimisation techniques, which have an ***iterative*** nature.

This includes gradient-based techniques as back-propagation and GA. Application of genetic algorithms for identification of *flexible* rule-based models will be considered later in Chapter 3.

✓ ***quasi-linear*** approach (data space clustering + linear least squares)

The second approach is to use clustering of the data space as a first step in order to define sub-spaces of interest and to use linear least squares to estimate the parameters of the consequent part. This approach is computationally very effective and will be used in Chapter 6.

As the TSK model is non-linear one, the second alternative (quasi-linear approach) leads to possible sub-optimality of the resulting model, which could affect to some extent the precision of the final model. As it is known, however, the precision and the transparency are related and a balance, a trade-off between them is more reasonable to be sought instead of ignoring either one.

In this context, there are works aiming to improve the precision together with the transparency (or refine the model) and the results are promising: significant improvement of the precision has been reported with simplification of the model structure in parallel (Setnes and Roubos, 2000; Angelov and Buswell, 2001b).

High computational efficiency of the *quasi-linear* approach is the reason to use it as a basis for the ***recursive*** approach, presented in Chapter 6. The ***non-iterative*** nature of the *mountain* and respectively *subtractive* clustering approaches makes them appropriate for *on-line* implementation.

The ***e***volving **R**ule-based (***e*R**) models and algorithms for control, fault detection and diagnostics, decision support and robotics, build using this *recursive on-line* identification scheme could be a basis, tool for design of *intelligent* or *smart* adaptive systems, the notion about which is presented later in Chapter 5.

4 NON-LINEAR APPROACH TO (*OFF-LINE*) IDENTIFICATION OF *FLEXIBLE* MODELS

The non-linear approach relies on numerical optimisation techniques like GA and gradient-based approaches. It should be mentioned that they are *iterative* and computationally more expensive.

The application of gradient-based techniques to such problems is often hampered by the specifics of the objective function such as absence of derivatives, non-continuity, non-convexity etc. Details about application of gradient-based neural network training procedures for *flexible* models identification could be found in (Jang, 1997).

GA, however, are well suited to the problem of structure and parameter identification because they are able to search complex, highly dimensioned spaces while being able to avoid local maximums and minima. Later in this chapter the application of GA for identification of *flexible* rule-based models is considered. The approach, which is applied, combines simultaneous structure and parameter identification.

4.1 Identification Problem Formulation

This combined (structure and parameter) identification problem could be formulated in the following form (Angelov, 2000):

To determine the fuzzy rules (represented by their indices) and their parameters such that to minimise the deviation between the model and the experimental outputs:

$$MSE = \frac{1}{N}\sum_{i=1}^{N}\left\| y - \hat{y} \right\|^2 \rightarrow \min \qquad (4.1)$$

subject to

flexible rule-based model described by (3.27)-(3.29)

$0 \leq Index \leq TNR; \qquad Index \in R^K$

$$(j-1)\frac{\overline{x}_l - \underline{x}_l}{m_l + 1} \le FLT_{lj} \le \frac{\overline{x}_l - \underline{x}_l}{m_l + 1}(j+1)$$

where $\underline{x}_l$ is the lower boundary of the lth linguistic variable for $l = \{1,2,...,(n+1)\}$;

$\overline{x}_l$ is its upper boundary; $j = \{1,2,...,m_l\}$;

N is number of training samples;

MSE denotes mean squared error;

FLT denotes *flexible* linguistic term.

It is important to note that the *flexible* model (3.27)-(3.29) is considered as one of the constraints.

4.1.1 Identification Criteria

An important element of every identification problem is the criteria (objective function). It defines how adequate the model is to the real process (object of modelling), represented normally by the experimental data.

The basic criteria is dated back to the XIXth century and has been formulated by Gauss while determining the orbit of the asteroid Ceres. It is called *the principle of least squares*. It could be expressed by minimisation of the sum of squares of the error determined by the deviation of the model outputs from the experimental data (4.1).

It should be mentioned that the least squares principle supposes a linear in parameters model. Extension for the case of stochastic disturbances (maximum likelihood method) has also been developed (Astrom and Wittenmark, 1990). For the more general non-linear case, least squares are also used as a good estimate, but different extensions have also been considered. One such extension has been tested in (Angelov et. al, 2000b), which combines *MSE* and amplitude of the error value:

$$E = \|y - \hat{y}\|^2 + \max | y - \hat{y} | \tag{4.2}$$

where $\dot{y}$ is the experimental output;

E - modified error function.

This results in some priority to averaging out the error so that there are fewer 'very poor' points at the expense of spreading the error more evenly over the whole range.

Additional improvement could be gained by adding a penalty over the smoothness of the model structure in the search criterion. In addition, convergence checks on the rule structure and estimated parameters could be incorporated to terminate the search without falling into over fitting.

4.2 GA - Brief Introduction

A very brief introduction to GA is given in this section in the sake of the completeness of the presentation. More details on binary encoded GA could be found in (Goldberg, 1989) and on real-value encoded GA in (Michalewicz and Fogel, 1999) as well as in the proceedings of the latest GECCO conferences (GECCO, 2001).

The main specific of the GA as an optimisation method is their implicit parallelism, which is combined with the evolution and hereditary-like process (Michalewicz and Fogel, 1999). GA is, in fact, a *driven stochastic search* technique, which combine stochastic (represented by *mutation* operator) and 'logical' search (represented by *crossover* of parental chromosomes and survival of the fittest by appropriate *selection*). GA probes a set of trial points (population) at each iteration called *epoch*.

Table 4.1. Population of Individual Chromosomes

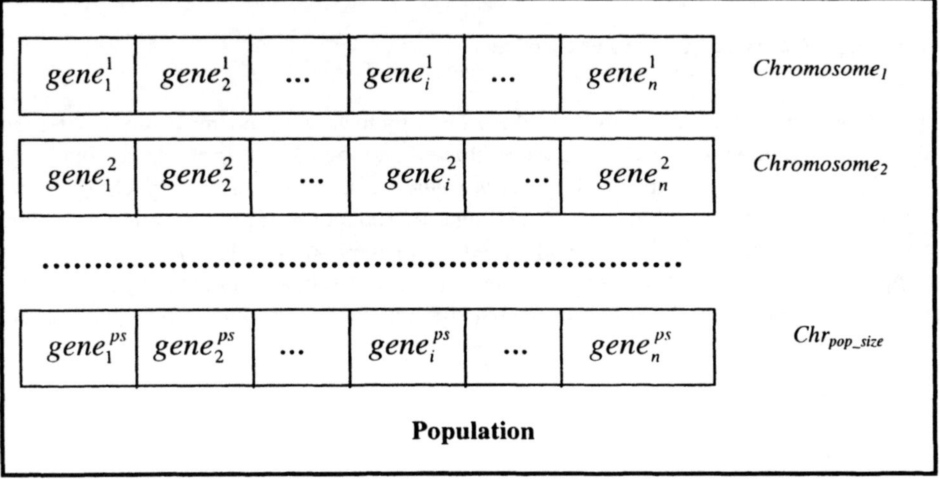

$gene_1^1$	$gene_2^1$	...	$gene_i^1$	...	$gene_n^1$	*Chromosome*$_1$
$gene_1^2$	$gene_2^2$	...	$gene_i^2$	...	$gene_n^2$	*Chromosome*$_2$
....						
$gene_1^{ps}$	$gene_2^{ps}$	...	$gene_i^{ps}$	...	$gene_n^{ps}$	Chr_{pop_size}

Population

The population consists of a number of chromosomes (Table 4.1). Each of the chromosomes represents a trial point in the search space. A gene in the chromosome represents a given problem variable.

In the binary-coded GA (Goldberg, 1989) each gene is represented by a number of bits each of them having value 0 or 1. For example:

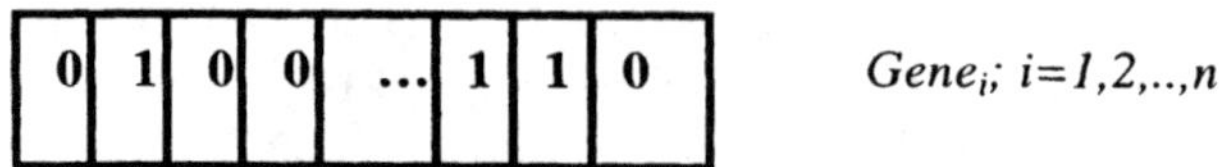

Fig. 4.1. Binary gene encoding

In real-coded GA (Michalewicz, 1996) each gene represents a single variable:

Fig. 4.2. Real-value chromosome encoding

Fitness is a function of the variables (x) which have to be maximised. At each epoch all individual *chromosomes* are evaluated and their *Ftness* calculated. Based on the *Fitness* values, some of the *chromosomes* from the current epoch (parental *chromosomes*) are selected for mating and reproduction. This operation is called **selection** for *mating*. The most popular **selection** method is called *elitist* **selection.** It supposes that the most fitted *chromosome* has a reserved place in the next population.

The following main operations are usually applied for producing new (child) trial points (*chromosomes*):

- **Selection**
- **Crossover**
- **Mutation**
- **Reproduction**

There exist various types of their realisation. The simplest *one-point* **crossover** could be represented as given in the Fig. 4.3.

For binary-coded GA **mutation** is a triggering from 0 to 1 and vice versa. For real-coded GA different schemes for **mutation** exist. The simplest one supposes generation of a random real number, which lays between the lower and upper boundary. The probability certain gene to be mutated (to change its value) is given by the probability of mutation.

Parental chromosomes (before mating):

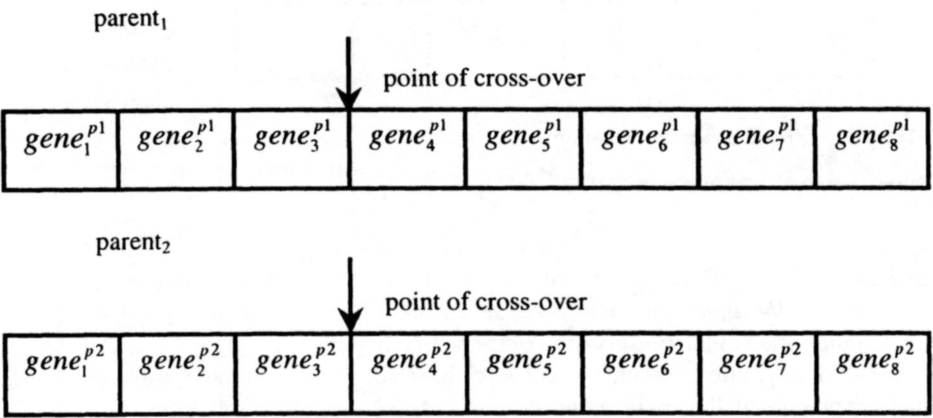

Child chromosomes

$Offspring_1$

$gene_1^{p1}$	$gene_2^{p1}$	$gene_3^{p1}$	$gene_4^{p2}$	$gene_5^{p2}$	$gene_6^{p2}$	$gene_7^{p2}$	$gene_8^{p2}$

$Offspring_2$

$gene_1^{p2}$	$gene_2^{p2}$	$gene_3^{p2}$	$gene_4^{p1}$	$gene_5^{p1}$	$gene_6^{p1}$	$gene_7^{p1}$	$gene_8^{p1}$

Fig. 4.3. Simple one-point crossover

where p_l denotes $parent_1$;
p_2 denotes $parent_2$;

The most commonly used mutation operator could be represented as:

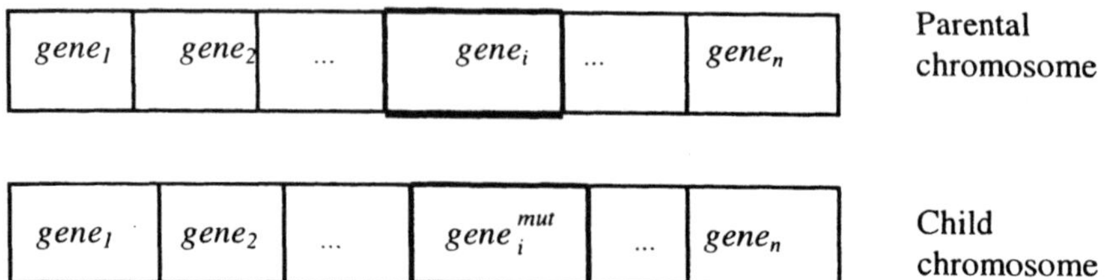

Fig. 4.4. Simple one-point mutation

where $gene_i^{mut}$ is a mutated gene.

Specifics of GA offers possibilities for their improvement by appropriate balance between *exploration* (possible because of the diversity in the population) and *exploitation* (due to the preservation of the search logic).

Historically, improvements of GA have been sought first in the optimal proportion and adaptation of the main parameters of the GA, namely probability of mutation, probability of crossover, population size (Grefenstette, 1986; Davis, 1989). More recently, the attention has been shifted to the **breeding** or **mating** (Muhlenbein and Schlierkamp-Voosen, 1993), i.e. to the process of forming new trial chromosomes at each epoch.

4.3 Centre-of-Gravity-based Crossover Operator

A special crossover operator based on the notion of the centre-of-gravity (CoG) has been introduced recently (Angelov, 2001). It performs specific breeding between the two fittest parental chromosomes. The new child offspring takes into account both the parental importance (weight, measured by their fitness) and their actual value.

Analytical proof of its ability to improve the result has been provided (Angelov, 2001) for the simplest case of one variable and when *elitist* selection strategy is used. This new operator has been validated with a number of usually used numerical test functions as well as with a practical example of supply air temperature and flow rate scheduling in a hollow core ventilated slab thermal storage system (Angelov and Wright, 2000).

The tests presented in the third part of the book indicate that it improves results (the speed of convergence as well as the final result) without practically increasing computational expenses.

The CoG-based crossover operator is more *informative* than mutation alone and more *innovative* than crossover itself. It increases diversity by creating a new chromosome different to the previous population elements and in the same time preserves the *search logic* by accumulating weighted information about parental population. It is designed to be used in addition to (in combination with) other crossover and mutation operators both in binary and real-valued GA.

It considers one of the child chromosomes to be produced by a special breeding of the two best fitted parental chromosomes (called $chromosome^{one}$ and $chromosome^{two}$), while the entire rest (pop_size-1) child chromosomes are produced in a usual way. One place in the population is preserved for this special chromosome, which represents the centre of gravity of $chromosome^{one}$ and $chromosome^{two}$ from the previous population (Fig. 4.5):

Chromosome_1^i
Chromosome_2^i
....
CoG^{i-1}
....
Chromosome_n^i

Fig. 4.5. Population Π_i

The resulting child chromosome is determined as CoG of the two parental chromosomes (Angelov, 2001):

$$CoG = \frac{chrmsm^{one} * Fit(chrmsm^{one}) + chrmsm^{two} * Fit(chrmsm^{two})}{Fit(chrmsm^{one}) + Fit(chrmsm^{two})} \tag{4.3}$$

where *chrmsm* denotes *chromosome*
Fit denotes *Fitness*

Every gene of the child chromosome is determined as a centre of gravity of the two parental ones:

$$gene_i^{CoG} = \frac{gene_i^{one}\, Fit(chrmsm^{one}) + gene_i^{two}\, Fit(chrmsm^{two})}{Fit(chrmsm^{one}) + Fit(chrmsm^{two})} \quad ;i=1,\ldots,n \tag{4.4}$$

where $gene^{one}$ denotes a gene from the $chrmsm^{one}$;
$gene^{two}$ denotes a gene from the $chrmsm^{two}$.

4.3.1 CoG-based Cross-over Operator - How It Works

Let us consider a simple

Example 4.1

It illustrates the CoG-based operator in action. A sample population consists of chromosomes having 6 real-coded genes each. Let the best chromosome at the ith epoch be:

$$chrmsm^{one} = [10;30;40;20;18;32] \tag{4.5}$$

Let its fitness is $Fit(chrmsm^{one}) = 0.7$. Further, let the $chrmsm^{two}$ and its fitness be respectively:

$$chrmsm^{two} = [18;16;15;13;12;22] \quad Fit(chrmsm^{two}) = 0.6 \tag{4.6}$$

Then the CoG-based child chromosome will be:

$$CoG = [13.7;23.5;28.5;16.8;6.6;27.4] \tag{4.7}$$

because

$$gene_1^{CoG} = \frac{10 * 0.7 + 18 * 0.6}{0.7 + 0.6} = 13.7;$$

$$gene_2^{CoG} = \frac{30 * 0.7 + 16 * 0.6}{0.7 + 0.6} = 23.5, \text{ etc.}$$

4.3.2 CoG-based Operator - Why It Works

Though, it is difficult to prove strongly that some new operator in GA is better even for some class of problems because of the probabilistic nature of the GA

(Michalewicz, 1996), it can be expected that in many cases CoG-based chromosome will have better *Fitness*.

This could easily be illustrated with the following very simple

Example 4.2

Let us consider the following one variable function $F(x)= e^{-0.1(\frac{x^2-7}{x})^2}$ (Fig.4.6).

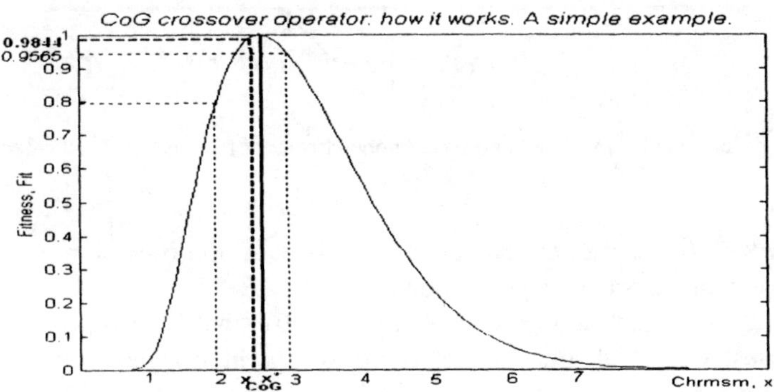

Fig. 4.6. How CoG-based crossover operator works - simple one variable example

Suppose, the ith population be $\prod_i = [1;2;3;4;5;6;7;8;9;10]$. The *chrmsm*^{one} and *chrmsm*^{two} will obviously be respectively:

$$\boldsymbol{chrmsm^{one} = 3;} \qquad Fit(chrmsm^{two}) = e^{-0.0444} = 0.9565 \qquad (4.8a)$$

$$chrmsm^{two} = 2; \qquad Fit(chrmsm^{two}) = e^{-0.225} = 0.7985 \qquad (4.8b)$$

The CoG-based new child chromosome/gene (in the case of one variable the chromosome is equivalent to the gene) then will be:

$$\boldsymbol{CoG^{i}} = \frac{2e^{-0.225} + 3e^{-0.0444}}{e^{-0.225} + e^{-0.444}} = \boldsymbol{2.455} \qquad (4.9a)$$

$$Fit(2.455) = \mathbf{0.9844} \tag{4.9b}$$

It is easy to see that it is much closer to the real maximum ($\sqrt{7} = 2.6458$).

In the next population *CoG*-based child chromosome is considered and, in addition (if *elitist* strategy is used), the best chromosome $chrmsm^{one}$ is also taken:

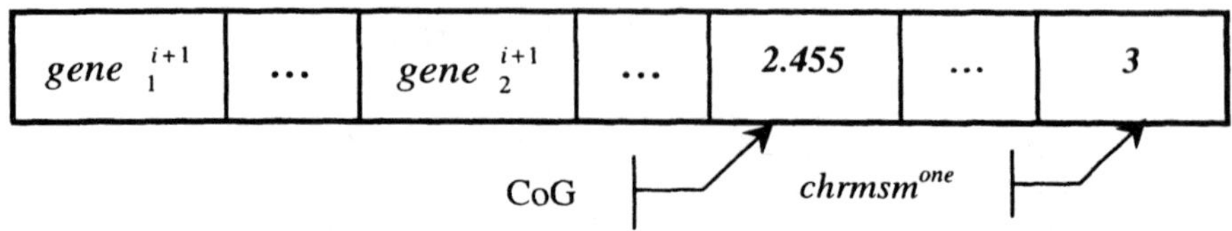

Fig. 4.7. Population (Π^{i+1}) with a CoG-based gene/chromosome (using elitist selection)

The other 8 chromosomes are produced by **crossover** and **mutation** of the parental chromosomes from the previous generation as usual.

Analytically it is possible to prove (Angelov, 2001) that improvements will occur for the simplest case with one variable and convex in the interval $(x_-;x_+)$ fitness function (*Fit*) when:

$$x_- \leq x^* \leq x_+ \tag{4.10}$$

where $x^* = \{x \mid Fit(x^*) = max(Fit)\}$

$x_- = min\ (crmsm^{one},\ chrmsm^{two})$

$x_+ = max\ (crmsm^{one},\ chrmsm^{two})$

Real situations, however, are more complex, but as the test results indicate improvements often occur. This could be explained with the fact that CoG-based child chromosome is produced by the two best parent individuals incorporating also information about their *Fitness*. By differ from the simple hill climbing it determines the new (often better) value of variables (x) directly (without using an estimation of the gradient and a step, which is usually computationally expensive, problem-dependent and a source of subjectivity).

4.3.3 Test Examples

The supplementary crossover operator has been tested with a number of commonly used in the literature test functions. A practical problem of scheduling of the supply air temperature and flow rate to a ventilated slab thermal storage system is also presented in the third part of the book.

4.4 Encoding and Decoding Indices of *Flexible* Rules and Linguistic Terms

Applying GA for non-linear identification of *flexible* rule-based models requires an appropriate encoding of model structure, namely *flexible* rules and membership functions' parameters into a chromosome. This is an important in terms of the efficiency of the identification problem solving procedure.

Simultaneous structure and parameter identification supposes larger number of unknowns and therefore, more compact, economical representation of the chromosome is required. Thus, encoding of the complete set of *flexible* rules as in earlier realisations of the non-linear approach (Bastian, 1996; Nelles, 1996) is time consuming and, in addition, the practical feasibility diminishes, as the number of rules becomes large.

It has been proposed to just encode the rules that participate in the *flexible* model (Angelov, 2000). The advantages are twofold:

- ✓ a significant reduction in the size of the chromosome;
- ✓ increased interpretability due to the reduction in the number of rules.

Different encoding and decoding procedures are possible. The principle requirement is that they have to be unique in both directions. Figs 4.8 - 4.10 depicts three possible chromosome configurations for a problem with five inputs ($n=5$), one output and a maximum of *7 fuzzy* linguistic terms.

The typical binary encoding is shown in the Fig. 4.8, where '0' means that the respective rule is not in the model and '1' means that it is in the model. Similar encoding mechanism is used for rules only (without parameters) in (Nelles, 1996). In the Fig. 4.8 membership function parameters are added as a second part of the chromosome.

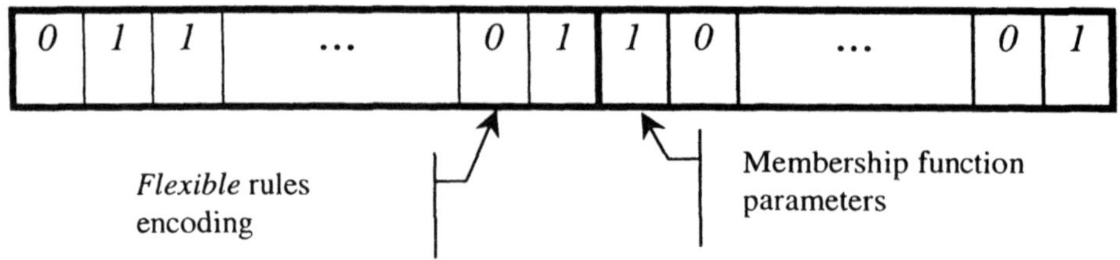

Fig. 4.8. Binary encoding using all possible rules

Encoding all possible rules in this way for the stated problem *(n=5; m=7)* results in a chromosome made up of a huge number of bits. This number increases further when the encoding of the parameters is accounted for.

Two new types of encoding based on rules indices and on linguistic label's indices has been proposed (Angelov et. al, 2000a). The advantage of encoding the rules in this way is that no information is lost, because the chromosome explicitly describes the rules.

By only considering part of the complete set of rules it is possible to reduce the length of the chromosome considerably. In the Fig. 4.9 it is shown how the chromosome length can be reduced to $k(n+1)$ integer (for encoded linguistic terms) and $2k$ real-coded genes (for membership parameters).

This size can be reduced further such that the length of the chromosome to be equivalent to k integer genes (representing the *flexible* rules' indices) + $2k$ real-valued genes representing the membership functions' parameters), by employing a base-encoding scheme, shown in the Fig. 4.10.

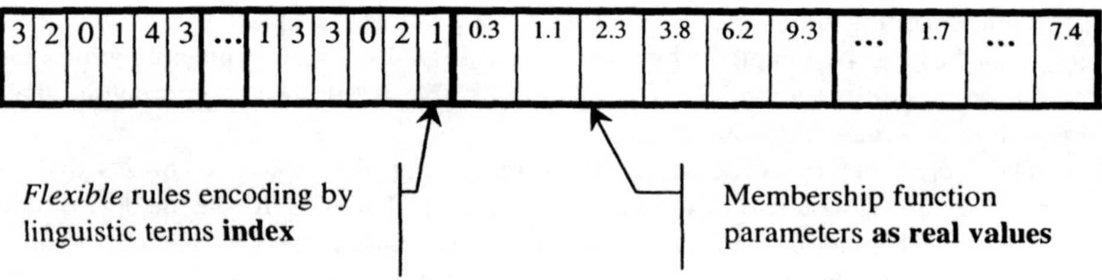

Fig. 4.9. Encoding based on linguistic term's index

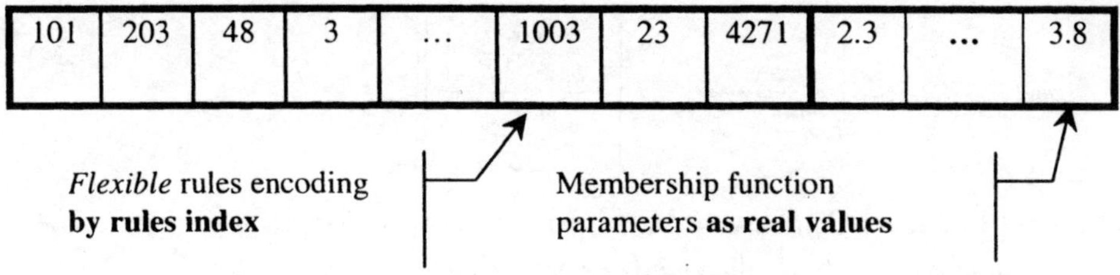

Fig. 4.10. Encoding based on rules' index

4.4.1 Encoding Procedure

It consists of two stages:

- The first stage sees each linguistic term translated into a *B*-based number; where $B = \max(m_i)\big|_{i=1}^{n+1}$ is the maximum number of linguistic terms.

  In this translation 0 could be assigned, for example, to lower linguistic term (e.g. *Very Low*), 1 to the next one (*Low*) and so on.
- The second stage transforms the *B*-based numbers (which represent the codes of the linguistic terms) to decimal integer positive numbers representing the index of the *flexible* rule:

$$Index = \sum_{i=1}^{n+1} Label_i base^{(n+1-i)} \tag{4.11}$$

where $Index$ is the decimal index of the respective rule;
$Label_i \in [0;(m_i - 1)]$ is the index of the linguistic term (it has normally linguistic meaning of a label)

4.4.2 Decoding a *Flexible* Rule

This is the reverse process, in which $Index$ is converted back into $(n+1)$, *B*-based numbers, calculated from (Angelov et. al, 2000):

$$Label_{(n+1+1-i)} = \left[\frac{Index_i}{base^{(i-1)}} \right] - T \left[\left[\frac{\left[\frac{Index_i}{base^{(i-1)}} \right]}{base} \right] \right]_{i=1}^{n+1} \qquad (4.12)$$

where $[.]$ indicates that only the integer part of the operation result is taken (this operation is often called *modulus* or *mod*).

Example 4.3

considers encoding and subsequent decoding of a *flexible* rule:

IF($x_1$is *Low*) **AND** (x_2 is *High*) **AND** (x_3 is *Very Low*) (4.13)

AND (x_4 is *Extremely High*) **AND** (x_5 is *Very Low*) **THEN** (y is *Low*)

where x_i are the input linguistic variables
y is the output linguistic variable.

Let the second input linguistic variable has five linguistic terms ($m_2 = 5$), namely:

- *Very Low* (index 0)
- *Low* (index 1)
- *Medium* (index 2)
- *High* (index 3)
- *Very High* (index 4)

Let, in the similar way, the other inputs and the output linguistic variables has seven linguistic terms ($m_{1,3,4,5 and 6} = 7$), namely:

- *Extremely Low* (index 0)
- *Very Low* (index 1)
- *Low* (index 2)
- *Medium* (index 3)
- *High* (index 4)
- *Very High* (index 5)
- *Extremely High* (index 6)

At the first stage, the codes of the used linguistic terms are determined, describing a specific rule. For the above example, the rule is represented by the following labels:

Labels = {2,3,1,6,1,2} (4.14)

Note that the "3" describing the *flexible* linguistic term associated with the second linguistic variable refers to *High* and not *Medium* because $m_2 = 5$. Selecting the base for the coding procedure gives, $base = \max\{7,5,7,7,7,7\} = 7$. Applying (4.11) to create the decimal integer ***rule' index*** yields:

$$Index = [(2*7^5)+(3*7^4)+(1*7^3)+(6*7^2)+(1*7^1)+(2*7^0)] = 41463,$$

Using (4.12) the decoding procedure can be shown to generate the following output,

$$i=1: \quad Label_{(6)} = \left\lfloor \frac{41463}{7^0} \right\rfloor - 7\left\lfloor \frac{\left\lfloor \frac{41463}{7^0} \right\rfloor}{7} \right\rfloor = 2,$$

$$i=2: \quad Label_{(5)} = \left\lfloor \frac{41463}{7^{-1}} \right\rfloor - 7\left\lfloor \frac{\frac{41463}{7^{-1}}}{7} \right\rfloor = 1,$$

and so on until,

$$i=6: \quad Label_{(1)} = \left\lfloor \frac{41463}{7^{-5}} \right\rfloor - 7\left\lfloor \frac{\frac{41463}{7^{-5}}}{7} \right\rfloor = 2.$$

If the number of *flexible* linguistic terms associated with each linguistic variable, (m_i), are equal the coding is most efficient. If the number varies, as in the above example, some redundancy in the indexes will be present. This may result in the GA occasionally selecting rule combinations that do not exist in terms of the problem definition. An advantage of the explicit rule coding method is that the crossover operator can be applied across either whole rules only *or* across any gene.

Both encoding mechanisms (using rules' indices and using linguistic terms' indices) were tested with real data and little difference observed between the two approaches in terms of convergence (Angelov et. al., 2000). Taking into account that encoding rules' indices leads to a smaller chromosome, it is computationally more efficient overall.

4.5 Algorithm of the Non-linear Approach

The basic algorithm applied to this problem could be represented by the pseudo-code given in the Fig. 4.11, where ε is the stop criterion (often a pre-defined MSE value) and "Generation" denotes number of generations.

On commencement of the search, the population of chromosomes is initialised randomly, but could be defined using *a priori* knowledge. These initial *parent* chromosomes produce *offspring* by application of genetic operators: selection, crossover and mutation.

Selection is performed for the whole chromosome because both parts contribute to the fitness value. The operator selects which chromosomes should remain in the population and sets the pairing for crossover. *Elitist* selection strategy in combination with the CoG-based cross-over is recommended.

```
Begin
    Generation = 0;
    Initialise (randomly or using a priori information)
    a population of chromosomes (Fig. 4.5);
    Decode fuzzy model by rules indices as in 4.12;
    Calculate outputs y using (3.27)-(3.29);
    While MSE ≥ ε
        Decode fuzzy model by rules indices as in 4.12;
        Calculate outputs y using (3.27)-(3.29);
        Evaluate Fitness;
        Perform (CoG) crossover and mutation;
        Perform selection and reproduction using Fitness
        Generation := Generation + 1;
    end
End
```

Fig.4.11. Pseudo-code of the GA used

Crossover and mutation are conducted separately on each part of the chromosome so there is no exchange of information between the parameters (represented by real values) and the structure (represented by integers). Crossover is applied at a randomly selected point along the length of paired chromosomes and the genetic material is swapped between them. In fact, crossover is the basic operator of GA as it is responsible for the productive interchange of the elementary information contained in the small building blocks.

At some given level of probability, genes in the chromosomes can mutate. Mutation operator is a background operator, which has important role about exploration capabilities of the algorithm. Mutated genes are replaced by a randomly generated integer or real number depending on which half of the chromosome the operation is applied.

The approach can be formulated to increase the flexibility of the search further. It is possible to incorporate additional parameters so the search for the optimal solution could be executed in terms of the number of linguistic terms associated with each variable, the membership function type, and the maximum number of rules desired (Angelov, 2000).

It should be noted, however, that additional controls would need to be incorporated into the objective function to control the increased model flexibility.

The flow-chart of information is represented in the Fig. 4.12.

4.6 Conclusion

The identification of *flexible* rule-based models in *off-line* mode is considered in this chapter. The basic phases of identification process as well as the different accents in the identification for control from the identification for other purposes has been outlined.

The existing approaches to identification of *flexible* rule-based models has been grouped in two basic types:

- ➢ Non-linear (based on numerical non-linear search methods);
- ➢ Quasi-linear (treating separately the antecedent and consequent part of *flexible* rules).

The identification is formulated as an optimisation problem and the specifics in criteria formulation has been considered. As the *flexible* models are non-linear and the least squares principle supposes a linear model, it is important to avoid over-fitting, to aim a smooth model structure (possibly by penalising distant rules), to take into account not just average error, but the absolute one as well etc.

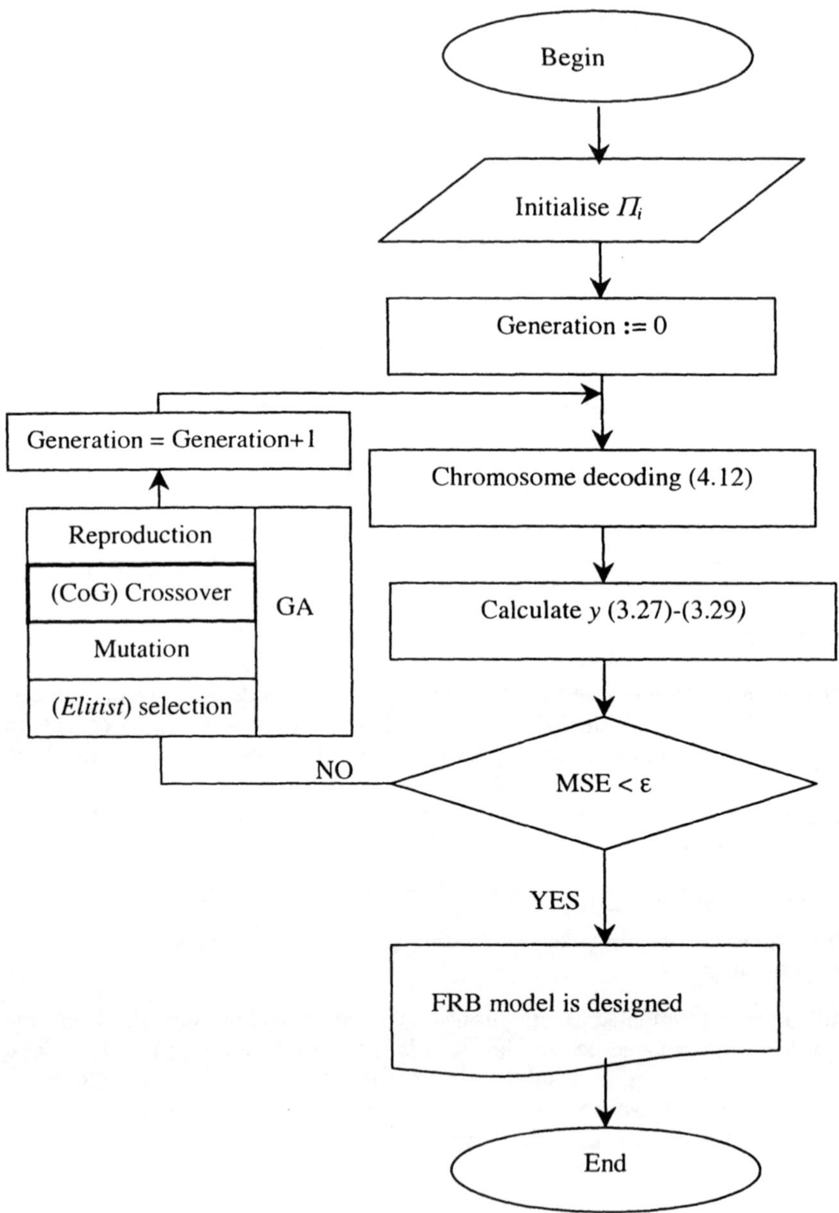

Fig. 4.12. Flow chart of the non-linear approach to flexible model identification using GA

The basic notion about GA has been briefly presented, as this numerical search procedure is beneficial with its ability to avoid local minima and its low requirements to the model structure (functions' derivatives, continuity, smoothness) and variables nature. An original crossover operator based on the centre-of-gravity paradigm has been also considered.

Treatment of the problem of proper encoding and decoding of the FRB models into chromosomes of the GA is presented. Two original types of encoding rules indices and linguistic terms into a two-part chromosome are represented and discussed. They are compared to the typical binary encoding in which huge number of genes is used (equal to the total number of possible rules).

Finally, the pseudo-code and flow-chart of the considered algorithm for non-linear identification of FRB models is presented at the end of this chapter. It has to be noted that this approach is applicable to all types of FRB models, including Mamdani-type and TSK models.

5 *QUASI-LINEAR* APPROACH TO FRB MODELS (*OFF-LINE*) IDENTIFICATION

The *quasi-linear* nature of TSK models allows separating the identification problem into two sub-problems:

- appropriate partitioning of the state space of interest by clustering;
- parameter identification of the consequent part.

As the output functions y_i are normally linear or singletons (constants), the second sub-problem is easy solvable by applying least squares technique (Astrom and Wittenmark, 1984).

For the first sub-problem clustering represents a more efficient alternative than the grid partitioning. The later is intuitively closer to the linguistic concept of *fuzzy* variables, but is impractical for larger dimensions, due to the so-called *curse of dimensionality*.

Mountain clustering and *subtractive* clustering are normally preferred than *fuzzy C-means* approach, mainly because of their non-iterative nature.

As a result of this separate treating of the identification problem, more effective computational algorithms is possible to be developed. As it will be demonstrated in the Chapter 7, the *non-iterative* nature of this approach makes it possible to develop *on-line* identification algorithms.

5.1 Data Space Clustering

The potential to be a cluster centre has been considered as a value to be optimised in the *mountain* clustering approach (Yager and Filev, 1993). This idea has been further developed in its modification called *subtractive clustering* (Chiu, 1994), in which every data point (sample) is considered as a candidate to be a cluster centre.

The potential is expressed as a sum of contributions inversely proportional to the Euclidean distances between a given point and all other data points, represented by the following formula:

$$P_j = \sum_{j=1}^{N} D_{ij} \tag{5.1}$$

$$D_{ij} = e^{-\xi\|z_i - z_j\|^2} \qquad i=1,2,\ldots N \qquad (5.2)$$

where P_i is the potential of the data point $z_i=[x_i, y_i]$ to be a cluster centre;
D_{ij} denotes the contribution of every single distance;
N is the number of training data samples;
$\xi = 4/r^2$; r is the cluster radii.

As it is clear from (5.1)-(5.2), the potential of a data point to be a cluster centre is higher, when more data points are closer to a specific candidate. The highest potential is called *reference* potential:

$$\overline{P} = \max_{i=1}^{N} P_i \qquad (5.3)$$

Example 5.1

It illustrates the notion of the potential in the case of a cooling coil (Fig.2.1) model with the following four inputs:

- ⇒ inlet water temperature (T_w); varies in the range [4;16], °C; 6 fixed values are considered;
- ⇒ inlet (to the coil) air temperature (T_a^{in});
  In fact this is the ambient air temperature; a *real* typical summer day (3 August 1998) is considered;
- ⇒ inlet air mass flow rate (m_a); varies in the range [0.2737;1.6424], kg/s; 6 fixed values are considered
- ⇒ control signal to the valve controlling the water mass flow rate (U_{cc}); varies in the range [0;1],-; 6 different values in this range are considered;

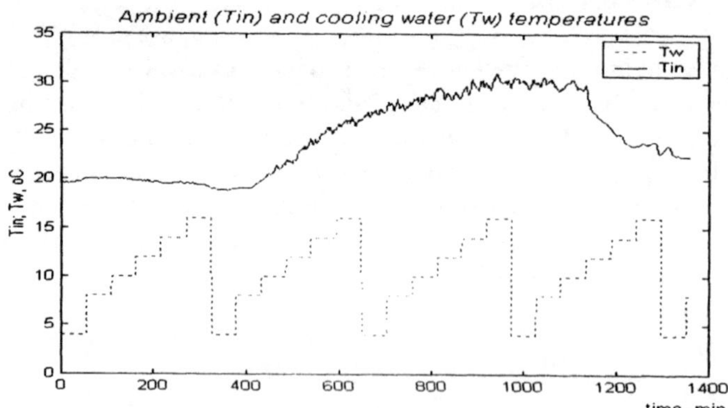

Fig.5.1. Ambient (T_a^{in}) and cooling water temperatures (T_w)

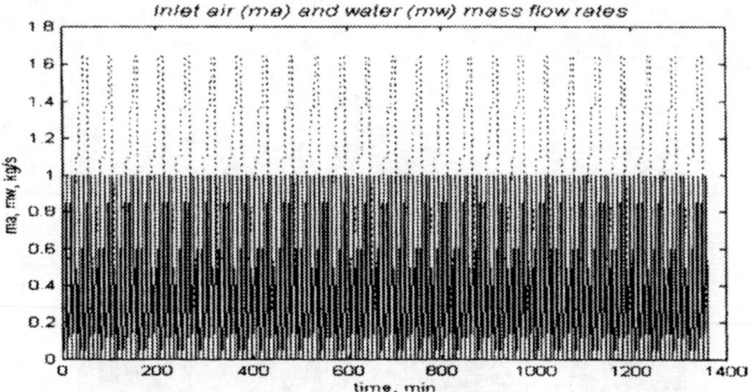

Fig.5.2. Inlet air (m_a) and water (m_w) mass flow rates

and the output:

⇒ outlet from the coil air temperature (T_a^{out})

In fact, the output represents the supply (to the zone) air temperature.

The sample data are represented in the Figs. 5.1-5.2 for the inputs and Fig. 5.3 for the output.

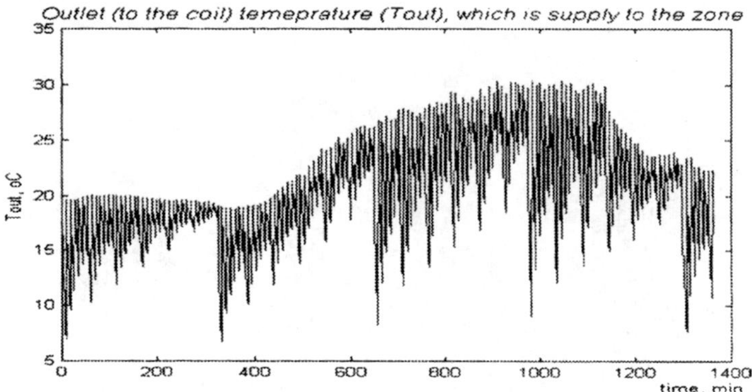

Fig.5.3. Outlet (from the coil) air temperature (T_a^{out}); supplied to the zone

The respective values of the potential of every data point (sample) are given in the Fig.5.4.

This example is considered in more details in the next part (III) of the book.

The expression for the potential (5.1)-(5.2) could be made independent on the number of the data samples by normalisation:

$$P_i = \frac{1}{N} \sum_{j=1}^{N} D_{ij} \qquad i=1,2,\ldots N \tag{5.4}$$

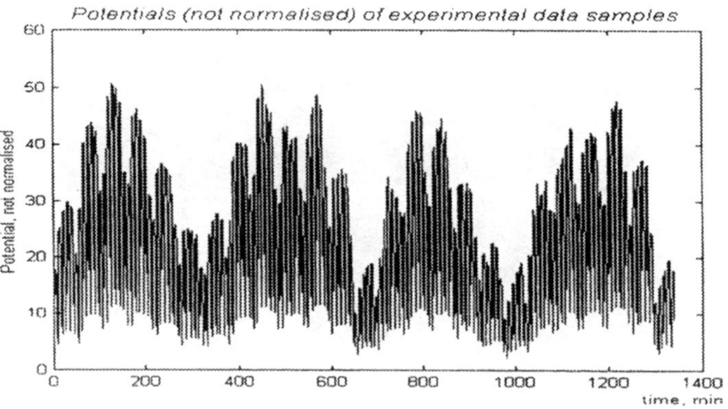

Fig.5.4. Potential of experimental data points

The normalised potential has value less than 1 (Fig. 5.5). As it will be seen later, the normalisation is important for *on-line* algorithms, because the number of considered data samples would grow up.

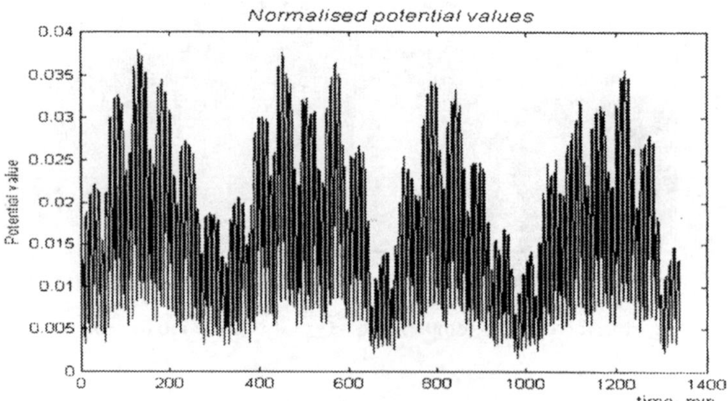

Fig.5.5. Normalised potential values for the Example 5.1

5.2
Subtractive Clustering

The procedure called *subtractive clustering* is based on the successive process of determination of the point with the highest potential (Chiu, 1994):

1. Initially, the point with the highest potential is chosen to be the first cluster centre.

2. Potential of all other points are then reduced with an amount proportional to the potential of the chosen point and inversely proportional to the distance to this centre:

$$P_i^{new} = P_i^{old} - P_k^* d_{ik} \quad i=1,2,...,N \tag{5.5}$$

where P_k^* denote the potential of the kth centre;
d_{ik} is a modified contribution inversely proportional to the distance to the already chosen centre (Chiu, 1994):

$$d_{ik} = e^{-\zeta \left\| z_i - z_k^* \right\|^2} \quad i=1,2,..., N \tag{5.6}$$

where z_k denote k^{th} cluster centre;
ζ is a positive constant (recommended value in (Chiu, 1994) is $\zeta=1.5\xi$)

3. A data point is chosen to be a new cluster centre, and respectively a centre of membership function, if its potential is higher than certain threshold, determined as a function of the reference potential.

4. If it is between a lower and upper thresholds, then additionally the shortest of the distances (δ_{min}) between the new candidate to be a cluster centre (z_k) and all previously found cluster centres is decisive. The following inequality, express the trade-off between the potential value and the closeness to the previous centres:

$$2P_m^* + \sqrt{\xi}\,\delta_{\min} P_1^* \geq 2P_1^* \tag{5.7}$$

This approach has been used for initial estimation of the *flexible* rules identification. It relies on the idea that each cluster centre is representative for a characteristic behaviour of the system (Chiu, 1994). The resulting cluster centres of the data space are used also as parameters of the *fuzzy* membership functions defining the rules of the model.

Typically, Gaussian type of membership functions has been used (5.8), but triangular ones are also a reasonable and often used option:

$$\mu(z) = e^{-\frac{(z^* - z)^2}{2\sigma^2}} \tag{5.8}$$

where z^* denotes the centre of the *flexible* rule
σ is the spread

5.3 Parameters (of the Consequent Part) Estimation

Second sub-problem (this of parameters of the consequent part estimation) is easily solved by applying linear least square technique (Astrom and Wittenmark, 1984). This is represented briefly below. The more general case of parameterised de-fuzzification (3.14) is considered. It is well known fact, that TSK model is linear by parameters. It is easy to demonstrate that it is the case also for the more general case of parameterised de-fuzzification.

BAD de-fuzzification gives additional leverage for trade-off between rules (giving higher priority to rules which have higher firing strength using higher value of the parameter $\theta\,(\theta=4\div5)$ and still considering all rules instead of mean of maximums approach). The model output y could be found as a linear combination of outputs of each rule y_i similarly to:

$$y=\sum_{i=1}^{R}\left(\sum_{j=1}^{n}\varpi_i(x)a_{ij}x_j+\varpi_i(x)b_i\right) \tag{5.9}$$

for

$$\varpi_i(x)=\frac{\sum_{i=1}^{R}\mu_i^{\theta}(x)x_i}{\sum_{i=1}^{R}\mu_i^{\theta}(x)} \tag{5.10}$$

Let $\Omega\in R^{NxN}$ denotes a diagonal matrix with $\varpi_i(x)$ as its diagonal elements. Let $\psi\in R^{Rx(n+1)}$ denotes vector of parameters of the consequent part of the TSK model. Finally, let $x_e\in R^{Nx(n+1)}$ is a matrix formed by addition of a unitary column to the vector of inputs of data samples.

Then parameters of the consequences could be found by linear least squares approach as the output is linear in respect to them:

$$\psi_i=\left[x_e^T\Omega_i x_e\right]^{-1}x_e^T\Omega_i y \qquad i=1,2,\ldots,R \tag{5.11}$$

Clustering of outputs could be used to determine singleton's values, if the outputs are constant (singletons).

5.4 *Flexible* Rule-based Model Refinements

5.4.1 Model Structure Simplification

The structure of the *flexible* model synthesised originally by the *quasi-linear* approach could be simplified based on the similarity of the respective linguistic terms. This is normally based on the respective membership functions (3.15).

In this sense, generalisation gains for the price of the precision, which, however, could be improved due to the ability of GA to escape local optimums using centres of the membership functions and parameters of the consequent part of the model as initial estimates only.

There could be close (and thus redundant) particular linguistic terms, while the rules' centres could be away of each other. It happens quite often in practice, as demonstrated in the Chapter 8.

The measure (5.7) for the closeness to the already existing cluster/rule centre (δ_{min}) concerns the rules as whole, but not the linguistic terms separately. Fig. 5.6 illustrates this for the case of two variables.

If suppose that the spreads of membership functions are the same (a very often used simplification), then the similarity between the respective *flexible* sets could be judged by the values of the centre parameters alone. An algorithm has been proposed (Angelov and Buswell, 2001b), which makes further simplification of the *flexible* rule-base structure based on the closeness of each particular linguistic term's membership function value.

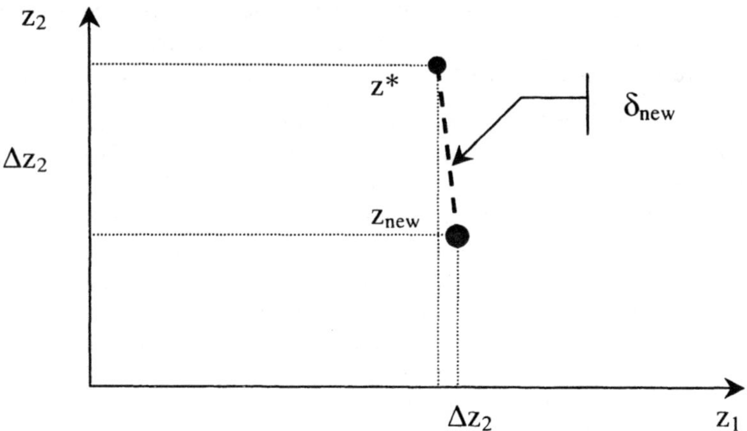

Fig. 5.6. Similarity on rule's centre and linguistic term's level;

where z^{new} denotes new data point/sample;
z^* is rule's centre;
Δz_1 denotes distance (in respect to the first variable) on linguistic term's level;
Δz_2 denotes distance (in respect to the second variable) on linguistic term's level.

As it will be demonstrated in the Chapter 8, it leads to significant simplification of the structure of the resulting *flexible* model without sacrificing the precision. This improves the interpretability of the resulting models, which is their main advantage comparing to the *black-box* models and is vital in applications such as fault detection and diagnostics, decision support systems and knowledge extraction.

The model structure could be further simplified and optimised and transparency, interpretation potential of the *flexible* rule-based model, could be improved by simplifying linguistic expressions using so called 'don't care' representation for specific linguistic variables.

5.4.2 Model Parameter's Refinement/Optimisation

As it has been mentioned the ***quasi-linear*** identification approach gives an (though quite good) ***approximate*** solution to the identification problem due to the separation of the original problem into two sub-problems. Several problems are possible to be solved only by application of the ***non-linear*** approach using *iterative* techniques, like GA or gradient-based techniques.

These includes:

- ✓ parameters of membership functions of the antecedent part identification;
- ✓ global optimum solution of the identification problem;
- ✓ refinement/optimisation of the model parameters.

As it has been illustrated (Roubos and Setnes, 1999; Angelov and Buswell, 2001b) it is possible to improve/optimise the model further using GA for parameter's refinement. It has been reported that the precision could be improved up to two times because of the ability of GA to find global optimums or to be close to them.

In this case clustering of data space and linear least square techniques are used as initial estimations only. Appropriate chromosome encoding could also ensure compactness with a pre-specified precision (Angelov, 2000).

The price for these improvements and *flexible* rule-based model refinements when GA are applied, however, is paid by the higher computational and time expenses which in control problems could have high importance by differ from classification problems. Therefore, we recommend to made this model refinements significantly rarely (at each K time sampling steps, $K>>1$).

5.5 Algorithm for (*Off-line*) *Quasi-linear* Identification of FRB Models

The (*off-line*) *quasi-linear* identification algorithm consists of the following basic stages:

- ✓ Stage 1 Determine the *flexible* rule-base structure by their centres and number of linguistic terms using data space clustering (applying subtractive clustering approach by Chiu (1994));
- ✓ Stage2 Refine model structure based on the similarity between the rules and linguistic terms;
- ✓ Stage3 Estimate parameters of the consequent part by linear least squares approach (or by clustering, if singletons are used);
- ✓ Stage 4 (optional) Refine parameters by non-linear optimisation (GA, gradient-based algorithm etc.).

The procedure of applying this approach is direct, *non-iterative* (except the optional last stage 4) and could be represented by the flow-chart depicted in the Fig.5.7.

5.6 Conclusion

The *quasi-linear* approach to identification of *flexible* rule-based models in *off-line* mode is considered in this chapter. It is based on the fact that TSK models are suitable to be treated in two stages:

- ➢ first the model structure represented by the *flexible* rules and centres of the membership functions to be determined by the clustering of the data space;
- ➢ next, the parameters of the consequent (linear) part of the models to be estimated in a *non-iterative* manner.

Data space clustering, which is in the core of this approach is considered first. The notion of the *potential* has been extended by normalisation, which is important for *on-line* realisation of the algorithm (as it will be demonstrated later in the book). An illustrative example of a cooling coil model is given.

Subtractive clustering is presented next, followed by estimation of parameters of the consequent part by linear least squares. The fact that TSK model is linear in the output by parameters, even if the BADD de-fuzzification is applied for the model output calculation, has been exploited.

Model structure simplification and parameters refinement is considered in the next section. The importance of this problem is due to the fact that the *quasi-linear* approach does not find the global optimum and, in fact, is an approximation to the original identification problem (4.1). But, as the investigation indicates, this model could easily be simplified and used as a very good initial guess for further improvement of the model precision and transparency using GA or *non-iterative* procedures.

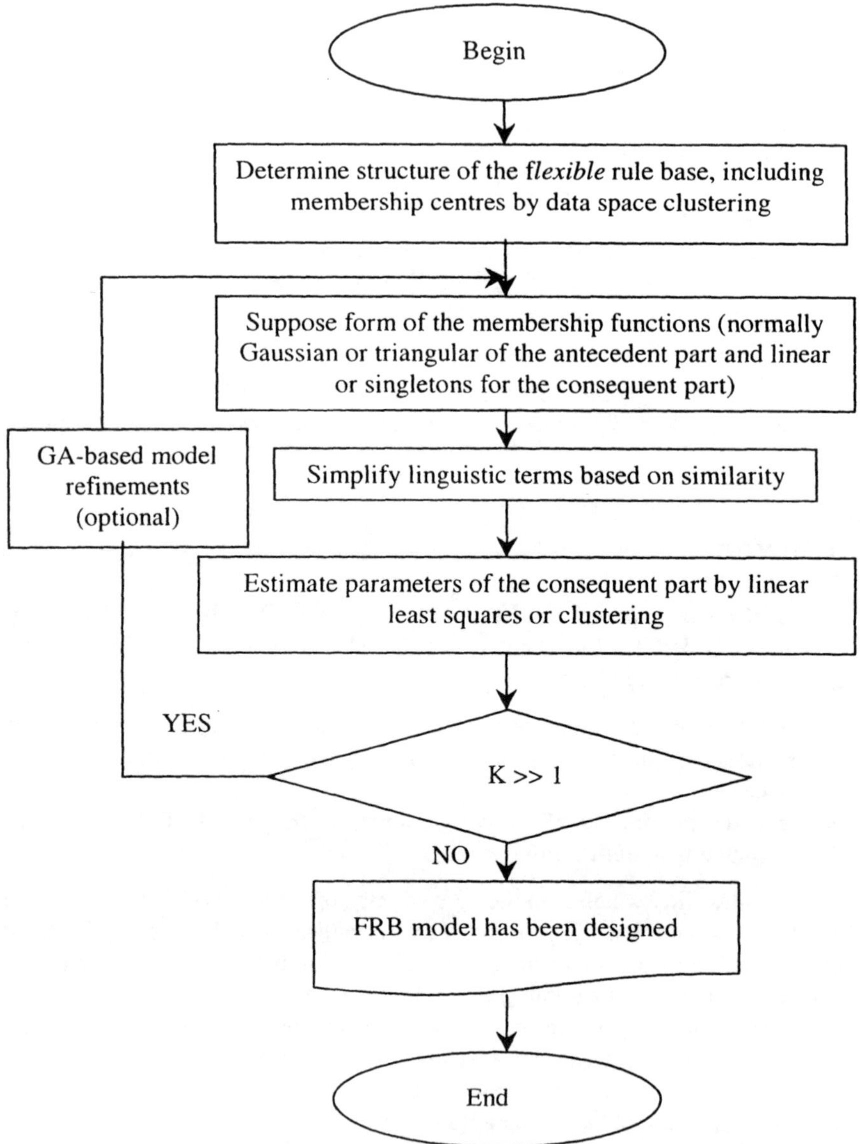

Fig. 5.7. Flow-chart of the *quasi-linear* approach of identification of FRB models

As it will be demonstrated later in the book, the *quasi-linear* approach could also be used as a basis of an *on-line* approach, appropriate for implementation in *real-time*.

6 *INTELLIGENT* AND *SMART* ADAPTIVE SYSTEMS

The ***recursive*** approach for *on-line* identification of *flexible* rule-based models is presented in more details in the next Chapter 7. The considered ***e***volving *flexible* **R**ule-based models (***e*R**) could be used as a tool for building *smart*, *intelligent* adaptive systems.

Necessity for such systems has been emphasised recently (EUNITE, 2000). They are a response of the science and theory to the pressing demands from different branches of industry and science to find an effective approach to adaptive yet *flexible* and robust representation, which to be computationally effective. They could be used in autonomous robotic systems, evolving and re-usable decision support systems, hardware, *smart* and *truly* (or, correctly, *more*) *intelligent* systems for fault detection and diagnostics, *on-line* control and parameter estimation, knowledge extraction and *intelligent agents* etc.

6.1 *Intelligent* Systems

6.1.1 Loose Definition

Generally speaking, *intelligent* systems (***i***-systems) suppose:

- ✓ capabilities for learning behaviour of the object of modelling or control;
- ✓ being able to take autonomous decisions, self-adapting and self-enriching their structure.

A more formal definition could be given as:

***i*-system is an adaptive *flexible* systems able to learn *on-line* behaviour of the object of modelling and control, to enrich, re-use and self-tune its structure and parameters.**

It have to be noted, that ***i***-systems are nowadays existing practice in areas like decision support, medical diagnostics, risk assessment, process control etc.

The area treating such systems is highly multidisciplinary. It is formed by the confluence and interaction of control theory and computational intelligence. More specifically, it includes knowledge extraction and data mining, behaviour-based modelling and control, *intelligent agents*, and other emerging branches.

6.1.2 Problems

Its main problems are:

- To find computationally effective precise, compact and transparent description of objects behaviour based on both quantitative and/or qualitative data available in *on-line* mode;
- To develop adaptive modelling and (optimal) control techniques based on such description;
- To bridge the gap between the theoretical advances and real engineering limitations in application of these techniques in practice;

6.1.3 Importance

This area has significant importance for making products and services more custom-oriented and *flexible*. Very often it leads to additional savings (energy, costs) by incorporating into the model factors, which has been previously ignored, possibly due to their qualitative nature.

i-systems normally surpass conventional ones, typically based on linear or simplified descriptions bounded to a number of assumptions, which are far from the real practice. Recent publications, including on the Internet, are emphasising its global importance. Internet itself has already employed many such customised and *intelligent* techniques in search engines, *intelligent agent* technology etc. (Bigus and Bigus, 1998).

For so called *intelligent* agents, which could be seen as a realisation (often software) of *smart* or *intelligent* systems an additional condition has place:

- they are normally part of a hierarchy trough a network of information and resource flows.

The main beneficiaries of the fast development of ***i***-systems are the large companies, providing more customised products and services, saving costs and energy and being ahead of the competition with *flexible* and *intelligent* solutions. Consumers themselves benefit by having higher quality of services and satisfaction of more *intelligent* products.

6.1.4 Specifics

A specific of this type of systems development is that they are a kind of *soft* technology: their implementation very often concerns basically the algorithmic part of already existing systems and networks and does not require costly hardware. Existing networks like LAN or Internet, existing PCs or microprocessors are very often used as the vehicles for their hardware and network realisation.

i-systems, which are, normally, based on application of *flexible* rules, neural networks or GA, have reached a stage close to wider practical application still lack, however, one very important property for a real-life sustainable and stable application: they are ***not adaptive*** in the full sense of this term, i.e. they are ***not recursive***, and they ***do not re-use*** the accumulated information during the process operation (EUNITE, 2000). While linear models and linear control theory as a whole has been developed to the stage to have effective solution to such requirements (Astrom and Wittenmark, 1984), this is not the case for complex, non-linear and linguistic models and systems.

The approach, presented in the next chapter is an example of combination of the *flexibility* of the non-linear, in general, rule-based models with the adaptive, recursive schemes, normally used in linear control theory by an effective mechanism of rules *innovation* and parameters up-date.

6.2 *Smart* Adaptive Systems

6.2.1 The Issue of *Smart* Adaptive Systems

The necessity of *smart* adaptive systems has been a matter of discussions very recently (EUNITE, 2000) as a result of development of the *intelligent* technologies (***i***-tech) which generally include a confluence of *fuzzy* systems, neural networks, evolutionary algorithms, pattern recognition and some related fields, like machine learning. During the last decade ***i***-tech became a viable and quickly expanding area of research and engineering practice.

Existing techniques and solutions, however, are normally ***not adaptive***, they did not allow ***re-use*** of the accumulated information during their generation. Sometimes they are called *self-learning* or *adaptive*, but they are rather self-adjusting and self-tuning. Their use in *on-line* mode and in *real-time* is possible for the cases when their structure is not changing.

Parameter adaptation, when possible, is time consuming as ***iterative*** training schemes are normally used. The last fact is due to the non-linearity of the considered models. In the same time, the lack of a true adaptivity hampers application of ***i***-tech, including their portability (the ability to use a system designed for one specific application to another quite similar one with as little modifications as possible).

6.2.2 Features of a *Smart* Adaptive System

To be really adaptive a system has to respond to possible changes in the environment (external influence/stimuli) or the object of modelling itself (possibly due to wearing, ageing, degradation or fault development). A desirable *intelligent* or *smart* adaptive system by differ from 'simply' adaptive (normally linear) systems have to be:

- ✓ able to change, evolve their structure simultaneously with their parameters;
- ✓ autonomous (to act and evolve on their own);
- ✓ able to react to a surprise, to unexpected input, to distinguish outlays;
- ✓ to accumulate experience (to build-up their structure during the routine operation);
- ✓ *intelligent* (to take decisions);
- ✓ adaptation mechanism itself could change (evolve) with time

Smart or *intelligent* adaptive systems is supposed to incorporate or mimic human-specific activities like perception, reasoning, and action to achieve multiple (possibly conflicting) goals while functioning autonomously in non-stationary environments.

6.2.3 Practical Implications

Examples of such systems exist in the nature and are represented with the living systems. Development of such technical systems (including software and Internet-based realisations) is highly demanded in different branches of social and economical activity.

An extremely simplified graphical representation of the idea of a *smart* adaptive system is given below. It should be noted that it does not intend to give a full representation, as it would be quite difficult, but rather to illustrate the idea.

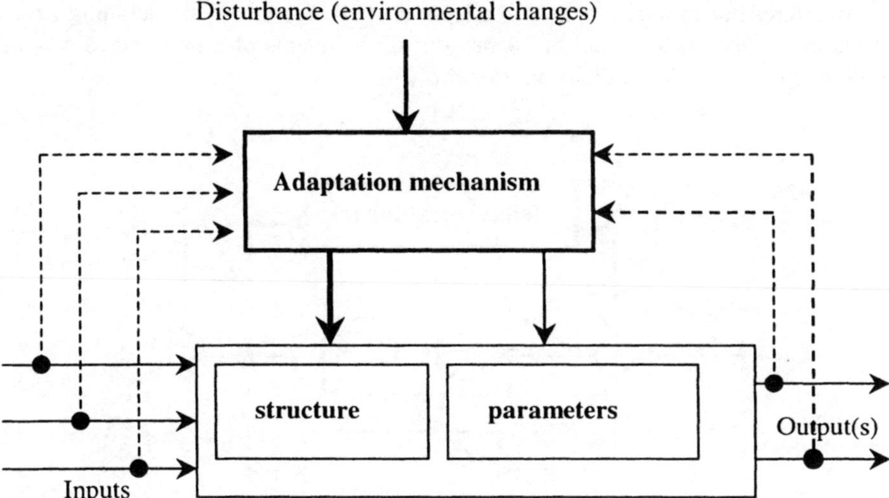

Fig. 6.1. *Smart* Adaptive System - basic idea

In this context, the approach for *on-line* recursive adaptation of rule-based models (called ***e*R** models), presented in details in the next chapter in details could be seen as a tool for building such *smart* adaptive systems. Systems, which are built using ***e*R** models, possess some of the features mentioned above. Some others (last two) could be implemented using them as a basis of a more complicated system.

6.2.4 *Intelligent* Indoor Climate Control System

There are specific fields of application in which the control object or the environment are changing significantly.

An example is a lecture theatre or an airport hall, where the occupancy pattern is significantly and constantly changing. If the object of modelling and control is the occupants' comfort, then obviously an adaptation is necessary. Moreover, parameter adaptation, which is normally meant when talking about adaptive control systems, could be non-adequate as the occupants themselves could go out and in freely. It is necessary to change/adapt also the structure of the model that describes their behaviour in addition to parameter adaptation in such cases.

An interesting application of such adaptive, evolving technique, including structure adaptation, could be to model the behaviour of occupants of a building as a basis of *intelligent* building air-conditioning systems (Fig. 6.2).

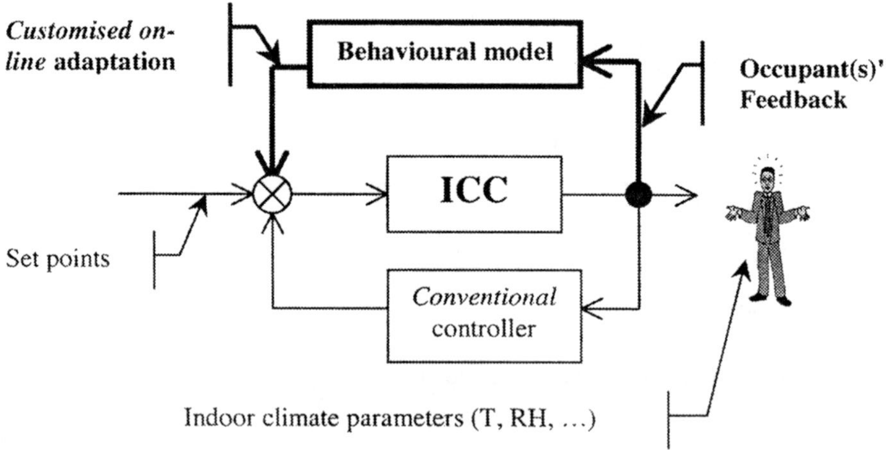

Fig. 6.2. *Intelligent* ICC System based on personalised behavioural model of **real** occupant(s)

Another possible application is an *intelligent* decision support system (DSS), which evolves and improves its performance taking into account new cases (new training data). A typical rule-based system is able to reason exclusively due to the pre-determined and fixed set of rules and thus can express no more decision-making ability than is explicitly contained within the knowledge-base. By differ from it, the ***e*R** models, represented in the next chapter 7, offer an ability to add, modify, analyse the rules generated in *real time*. In this way they make possible to reflect the changes in the environment or the object itself.

6.3 Conclusion

The notion about *intelligent* systems and its special case of *smart* adaptive systems have been considered in this chapter. Definition, problems and specifics of *intelligent* systems have been outlined. Their importance also has been mentioned. A definition and features of *smart* adaptive systems has been represented together with an illustrative schematic diagram. An idea about *intelligent* indoor climate control system has been represented and illustrated.

The ***e***volving **R**ule-based (***e*R**) models, which are considered later in this part of the book, could be seen as a tool for building such *smart* adaptive systems.

7 *ON-LINE* IDENTIFICATION OF *FLEXIBLE* TSK-TYPE MODELS

In this chapter a novel approach to *on-line* data-driven identification of *flexible* rule-based models is considered. It concerns primarily TSK *flexible* models benefiting from their dual nature: being non-linear, they are *quasi-linear* and, therefore, convenient for using in *on-line* identification schemes.

7.1 The Concept

The basic aim in developing this approach has been to build a simple, computationally effective mechanism for adaptation of *flexible* models. It has been aimed that this mechanism would allow enriching the structure of the model, instead of simply adjusting its parameters, as has been done very often. Therefore, these models could realise a real *'learning trough experience'*, which is potentially very useful in robotic systems, for fault detection and diagnostics etc.

The main features of this approach was aimed to be:

- ✓ To adapt, modify, *innovate* not only parameters of the model, but also its structure. In the same time this should happen not very often to avoid overloading and over-fitting redundancy;
- ✓ To enrich the structure of the model (not just to modify, but to add, if necessary, a building block (rule));
- ✓ To be as simple as possible (the structure to be as minimal as possible, to apply *non-iterative* techniques, when possible; suppose linear relations where admissible) even for the price of losing the global optimum value in identification and therefore to some extent the precision;
- ✓ To be evolving. It does not mean necessarily to use so called evolutionary algorithms, including genetic algorithms. It rather means to assume a growing structure of the model and ability to build-up it during the normal process of operation.

This property is particularly appropriate for robotics and fault detection and diagnostics, commissioning and *on-line* energy analysis.

7.2 Basic Phases of the Procedure

A *non-iterative*, recursive procedure implemented in *real-time* in *on-line* mode has been developed (Angelov and Buswell, 2001a) which includes the following basic phases:

✓ *Phase 1*

Start with a structure (rule-base) and membership functions parameters determined *off-line*. It could be only one rule. Few data samples could be used to feed the *off-line* algorithm for identification as described in the Chapter 5.

✓ *Phase 2*

Collect new data sample in *on-line* mode and do the next phases in *real time*. It supposes measuring, filtering and storing the data (if needed).

✓ *Phase 3*

Calculate potential of the new data sample in respect to all already collected data samples.

✓ *Phase 4*

Up-date *recursively* the potential of already existing *flexible* rules (and respectively clusters) centres taking into account the new data point's influence.

✓ *Phase 5*

Up-date the value of the reference (maximal) potential taking into account both the potential of the new data point and its influence to the other potentials.

✓ *Phase 6*

Rule-base *innovation* (including rules structure up-grade and modification). This is the basic phase and will be considered separately in the next section. Generally, at this phase the structure of the *flexible* rule-based model has been revised, modified or up-graded in dependence on the informative potential the new data sample brings and how it stands in comparison to the potential of already existing rules' centres. This is the key mechanism, which makes difference comparing to just parameter adaptation, adjustment or tuning, making possible building autonomous, *smart* or *intelligent* systems, which ***develop their structure*** during the normal process operation.

✓ *Phase 7*

Model structure simplification based on the similarity measures and closeness to the already existing centres and linguistic labels. This is done ***non-iteratively***.

✓ *Phase 8*

Recursive parameter estimation of the consequent part of FRB models. This phase is also considered in a separate section below.

Phase 1 is based on the *off-line* identification or an initial guess, which could be provided using existing expert knowledge (if available). It is very important to mention that due to the good adaptation properties of the algorithm, the requirements to the initial FRB model structure are not stringent. This is particularly important for implementations of the algorithm in robotics, *on-line* fault detection and diagnostics etc.

Phase 2 concerns mainly the hardware realisation and data acquisition process, which will not be considered here, as they are mostly problem-specific. For more details see e.g. (Astrom and Wittenmark, 1990).

7.3 Potentials Up-date in *On-line* Mode

After the *on-line* process starts, the informative potential of every new data, collected according to the conditions specified in the Phase 2, is calculated in *real-time* at Phase 3:

$$P^{new} = [P_{N+k+1}]^{k+1} = \frac{1}{N+k+1} \sum_{j=1}^{N+k+1} D_j \tag{7.1a}$$

$$D_j = e^{-\xi \|z_{N+k+1} - z_j\|^2} \qquad i=1,2,\dots N \tag{7.1b}$$

where $k=0,1,2,\dots$ denotes the time steps in *on-line* mode ($k=0$ represents the time instance when *off-line* mode finishes);

P^{new} denotes the potential of the new data sample

It has to be noted that in *on-line* mode the number of data samples is continuously growing up. Practically, it is not critical to have thousands of data samples or more, because only new data potential calculation (7.1a)-(7.1b) needs old data samples. The potentials of the data samples calculated in *on-line* mode (after the vertical dashed line) and these of the points calculated in *off-line* mode (first 400min; up to the vertical dashed line) are presented in the Fig.7.1 for the cooling coil considered in the Example 5.1.

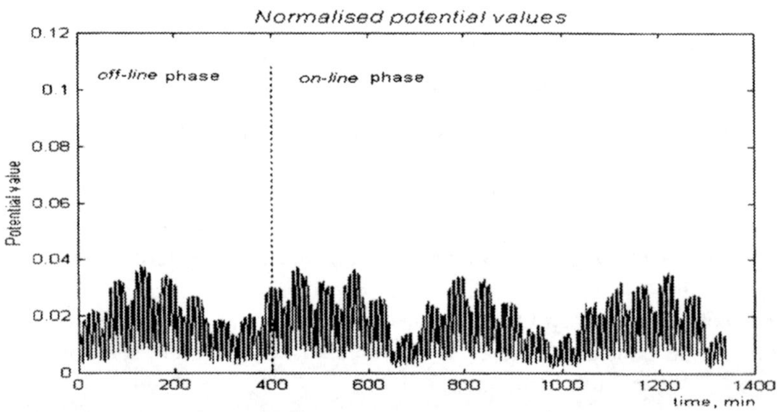

Fig.7.1. Potentials of *on-line* collected data point/samples and these of *off-line* collected ones (the vertical dashed line represents the instance when *on-line* mode starts)

As it is considered later, **only** the potentials of **old centres** (not all data samples) are up-dated. Output parameters of the Takagi-Sugeno model are calculated ***recursively*** using compressed old information, and the algorithm is a ***non-iterative*** one.

In order to avoid overloading memory in respect to the data (z), however, a *moving window* of consideration with length L has been introduced. Therefore, (7.1a)-(7.1b) is performed for $k=0,1,2...,\ L\text{-}N\text{-}1$, i.e. until the total number of data samples reaches L.

Practically (as the tests shows), this could be a large number (thousands data point or even more, because of the ***non-iterative*** nature of the algorithm) and depends on hardware and time restrictions. After this limit is reached, the following *sliding window* horizon is considered:

$$P^{new} = P_{N+k+1}\big|_{k+1} = \frac{1}{L}\sum_{j=N+k+1-L}^{N+k+1} D_j \quad ; k = L\text{-}N,\ L\text{-}N+1,\ldots \tag{7.2}$$

$$D_j = e^{-\xi\left\|z_{N+K+1}-z_j\right\|^2} \qquad i=1,2,\ldots N \tag{7.3}$$

At the next Phase 4 potentials of old centres are up-dated by adding the amount respective to the influence of the new data point in a ***recursive*** manner:

$$P_j^*\big|^{k+1} = \frac{N+k}{N+k+1} P_j^*\big|^{k} + \delta P_j^*\big|^{k+1} \quad ; j = 1,2,\ldots,R \tag{7.4}$$

$$\delta P_j^*\big|^{k+1} = \frac{D_j}{N+k+1} \qquad k=0,1,2,\ldots \tag{7.5}$$

where $P_j^*\big|^{k+1}$ denotes the potential of the $j^{\text{-}th}$ centre at the moment $k+1$ in *on-line* mode;

$P_j^*\big|^{0}$ is the potential of the $j^{\text{-}th}$ centre in *off-line* mode ($k=0$), which is already calculated;

$\delta P_j^*\big|^{k+1}$ denotes the correction to the potential of a rule' centre

$D_j = e^{-\xi\left\|z_{N+k+1}-z_j^*\right\|^2}$ denotes the contribution of each distance to a rule' centre from a new data point/sample

Time and computation savings are realised due to the use of already calculated potentials of clusters, which suppose calculations over large matrices (normally, the number of data samples is big: $N>10^3$).

As a next step, at Phase 5, reference potential (the maximal one) is also updated:

$$\overline{P}_{N+k+1} = \max(P_j^*, P^{new}) \quad j=1,2,\ldots,R \tag{7.6}$$

The reference potential could, generally, be affected due to:

- ✓ Correction (up-date) of already calculated potentials of centres. This corrections are different for each data point and their ranking could change;
- ✓ Appearance of a new data point with higher potential. Theoretically, the new data point (a possible new cluster/rule centre) could have bigger potential. It has to be noted, however, that this option is possible, only when a sliding window is considered and some data are forgotten. The reason is that when all data are considered the potential tends to have smaller value as it is normalised and the points are more scattered when new points are added (see the upper dashed line in the Fig. 7.3).

7.4 Rule-base *Innovation* and Modification Mechanism

This is the key process in FRB model adaptation, which makes the difference from the traditional adaptation (normally concerning only parameters). It has the potential to be the basis of building *smart* or *intelligent* adaptive systems, which ***evolve*** in response to the change in the environmental conditions or internal stimuli.

The mechanism for rule-base up-date, which we call rule-base *innovation,* is quite simple. This makes possible to build a computationally efficient algorithm based exclusively on the notion of data samples potential as a measure of their informative importance.

As a note, it could be mentioned that in future works it will be investigated the necessity and efficiency of considering some additional criteria, which could have higher priority, like safety rules or ignoring some data. This, in effect, would lead to a hybrid scheme of combination of data-driven technique with some knowledge-based constraints. At present, 'purely' data-driven approach is followed.

The rules *innovation* mechanism, which is potential-based, is illustrated in the Fig. 7.2.

First, the potential of the new data sample is compared to the potential of **existing centres** and, if it is greater than the potential of any other centre, a **new rule** is formed and is added to the rule-base. In this way the useful information contained in rule-base structure is preserved as well as unnecessary calculations over large matrices are avoided.

The centre of the new cluster and respectively the new rule is simply the new data sample, when its potential is enough to replace an old rule. It should be noted, that outlays between new data samples are automatically rejected, as their potential is significantly lower, because there is no many points surrounding them. This serves as a natural filter, which could help to distinguish false alarm from a real fault in fault detection and diagnostics.

If the potential of the new data sample is big enough ($P^{new} > \varepsilon P$, where ε stands for the upper threshold level) and it is not too close to previously found cluster centres, then the **new data is added as a new cluster** (and respectively new rule's) **centre**. This process is called ***rule base innovation***.

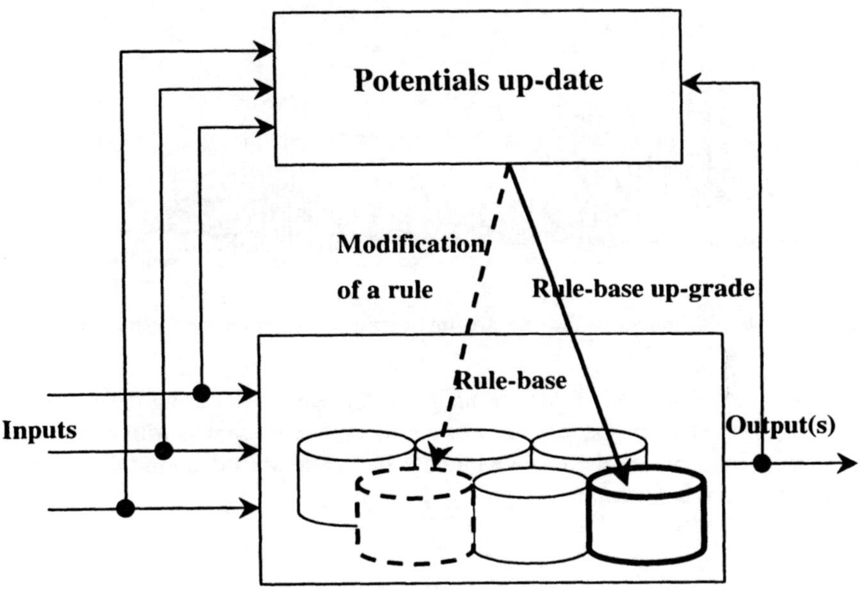

Fig. 7.2. Rules *innovation* and modification mechanisms

If it is higher than $\underline{\varepsilon}\overline{P}$ and higher than at least one of the previously accepted centres, and the new data sample is not too close to them, then it **replaces** the rule with the lowest potential. This process is called rules **modification**.

The *on-line* up-date of the rules structure using these reference lines (threshold levels $\underline{\varepsilon}$ and $\overline{\varepsilon}$ are multiplied by the reference potential value $\overline{P}$) is presented graphically by dashed lines in the Fig. 7.3 for the problem represented in the Example

5.1. It is seen, that new rules are added based on data samples from first 50 and around the 700th in *on-line* mode (see arrows in the Fig. 7.3).

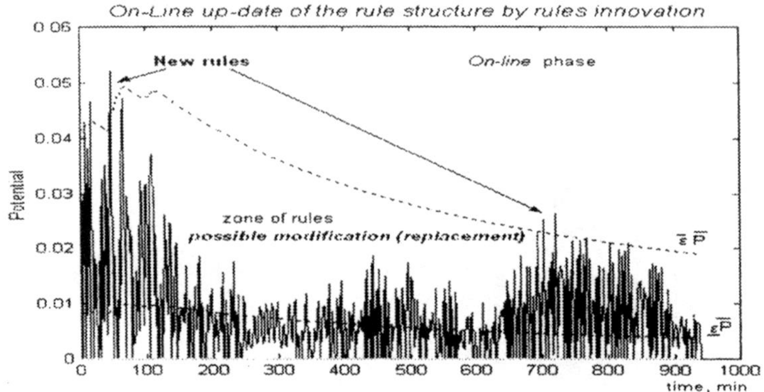

Fig.7.3. *On-line* up-date of the rule structure by rules *innovation* and modification

In the Fig.7.4 the rule structure up-grade by new rules during the process run is illustrated. Potentials of *flexible* rules of the model are presented with circles denoting the existing rules before the *on-line* up-date starts and stars denoting the *on-line* rules modification and *innovation*.

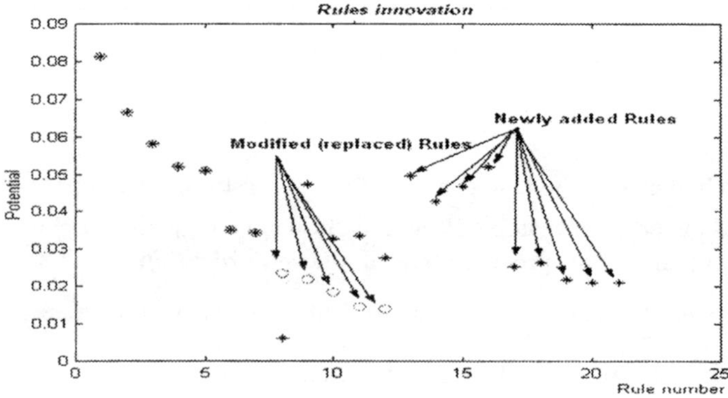

Fig.7.4. *On-line* up-grade (*innovation* and modification) of the rule structure

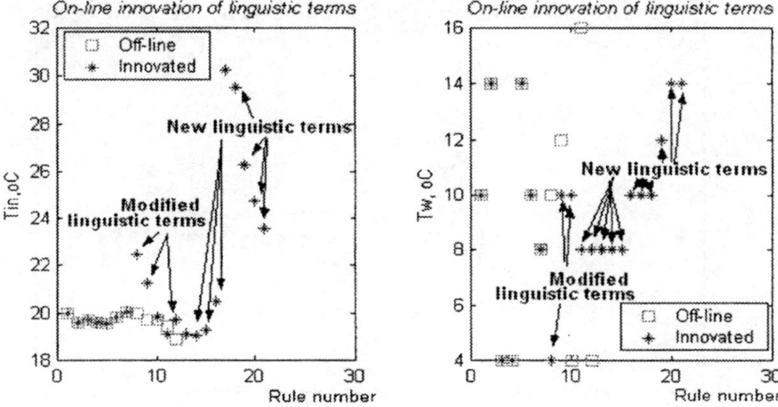

Fig.7.5. *On-line innovation* and modification of linguistic terms of T_a^{in}, T_w

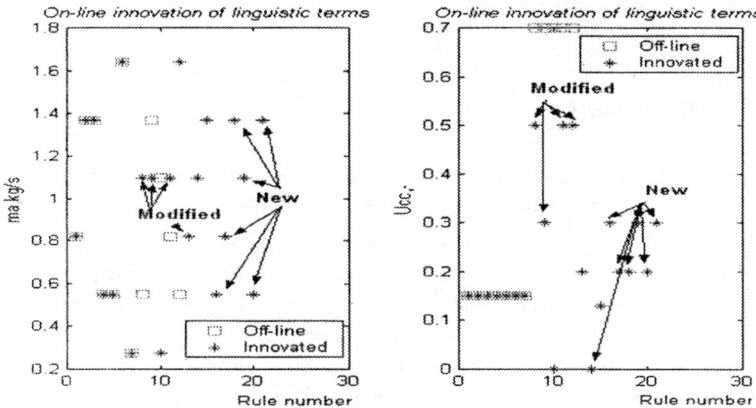

Fig.7.6. *On-line innovation* and modification of linguistic terms of m_a, U_{cc}

This up-grade of rules means that respective new linguistic terms are added for each *flexible* variable. For the Example 5.1 this process is illustrated in the Figs. 7.5-7.7, where for each input variable (T_a^{in}, T_w, m_a, U_{cc}) and the output variable (T_a^{out}) squares denote centres of respective linguistic terms before the *on-line* model up-date starts. The stars denote the respective modifications and the new linguistic terms added for each variable.

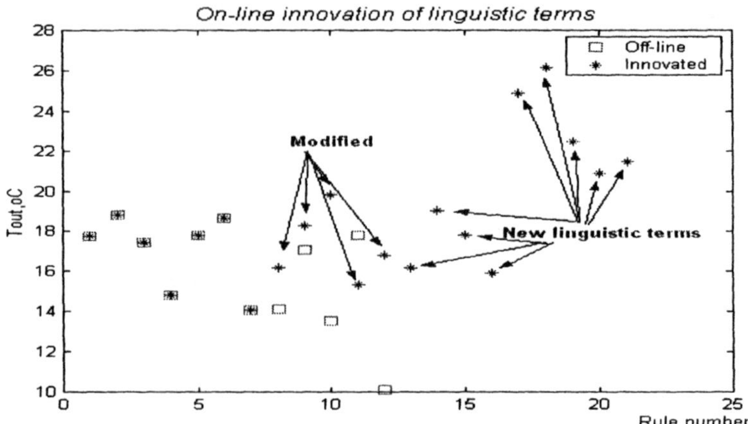

Fig.7.7. *On-line innovation* and modification of linguistic terms of T_a^{out}

7.5 Parameters Up-date

Parameters of the consequent part of the FRB model are estimated by linear least squares. The procedure for this estimation makes use of already calculated matrices.

Let the $\Omega \in R^{(N+k+1)(N+k+1)}$ denotes a diagonal matrix with $\varpi_i(x)$ as its diagonal elements. Let $\psi \in R^{R(n+1)}$ denotes vector of parameters of the consequent part of the TSK model. And, finally, let $x_e \in R^{(N+k+1)(n+1)}$ is a matrix formed by addition of a unitary column to the vector of inputs of data samples.

Then parameters of the consequences could be found by linear least squares approach (Astrom and Wittenmark, 1990) as the output is linear in respect to them:

$$\psi_i = [x_e^T \Omega_i x_e]^{-1} x_e^T \Omega_i y \qquad i=1,2,\dots,R \qquad (7.7)$$

Parameters of the consequence for this new rule are determined by the following

Procedure:

Step 1 From (7.7) vector of parameters of the consequent part is determined for the moment k in *on-line* mode:

$$\psi_i^{N+k+1} = [F_i^{N+k+1} x_e]^{-1} F_i^{N+k+1} y \qquad i=1,2,\dots,R \qquad (7.8)$$

where $F_i^{N+k+1} = (x_e^{N+k+1})^T\ \Omega_i^{N+k+1}$; $F \in R^{(n+1)*(N+k+1)}$

Step 2 Update **recursively** F by:

$$F_i^{N+k+1} = \left[\begin{array}{c|c} F_i^{N+k}\Psi^{N+k}\Xi_i^{N+k} & 0 \\ \hline 0 & \dfrac{\mu_i^\theta(x_{N+k+1})}{\sum_{j=1}^{R}\mu_j^\theta(x_{N+k+1})} \end{array}\right] \tag{7.9}$$

where Ψ and Ξ are diagonal matrices with elements in the main diagonal determined by:

$$\psi_i^{jj} = \begin{cases} \dfrac{\sum_{i=1}^{R_{before}}\mu_i^\theta(x_j)}{\sum_{i=1}^{R_{before}}\mu_i^\theta(x_j)+\mu_{new}^\theta(x_j)}; & Phase6 \\ 1; & otherwise \end{cases} ;j=1,2,..N+k+1; i=1,2,..,R \tag{7.10}$$

$$\xi_{new}^{jj} = \begin{cases} \dfrac{\mu_{new}^\theta(x_j)}{\mu_i^\theta(x_j)}; & Phase.6; \forall i=1,2,...,R_{before} \\ 1; & otherwise \end{cases}$$

$$\xi_i^{jj} = 1; i=1,2,...,R_{before}$$

Phase 6 denotes cases when Phase 6 of the procedure is activated (when a new rule is added or a rule is replaced by another one);
R_{before} denotes the number of rules before the update;
$\cdot_{new}$ is the new rule (the one which is added or replaces another one).

Step 3 Update vector of parameters (at the next moment $k+1$, when a new data sample is available) taking into account the growing data set by:

$$\psi_i^{N+k+1} = [Q^{N+k+1}]^{-1}S^{N+k+1} \quad i=1,2,...,R; \quad k=0,1,... \tag{7.11}$$

where $Q^{N+k+1} = Q_i^{N+k} + \Delta Q_i^{N+k+1}$

$$S^{N+k+1} = S_i^{N+k} + \Delta S_i^{N+k+1}$$

$$\Delta Q_i^{N+k+1} = \begin{pmatrix} x_{(N+k+1)1}\varpi_{N+k+1}x_{(N+k+1)1} & \cdots & x_{(N+k+1)1}\varpi_{N+k+1}x_{(N+k+1)n} & x_{(N+k+1)1}\varpi_{N+k+1} \\ x_{(N+k+1)2}\varpi_{N+k+1}x_{(N+k+1)1} & \cdots & x_{(N+k+1)2}\varpi_{N+k+1}x_{(N+k+1)n} & x_{(N+k+1)2}\varpi_{N+k+1} \\ \cdots & \cdots & \cdots & \cdots \\ x_{(N+k+1)n}\varpi_{N+k+1}x_{(N+k+1)1} & \cdots & x_{(N+k+1)n}\varpi_{N+k+1}x_{(N+k+1)n} & x_{(N+k+1)n}\varpi_{N+k+1} \\ \varpi_{N+k+1}x_{(N+k+1)1} & \cdots & \varpi_{N+k+1}x_{(N+k+1)n} & \varpi_{N+k+1} \end{pmatrix}$$

$$\Delta S_i^{N+k+1} = \begin{pmatrix} x_{(N+k+1)1}\varpi_{N+k+1}y_{N+k+1} \\ x_{(N+k+1)2}\varpi_{N+k+1}y_{N+k+1} \\ \cdots \\ x_{(N+k+1)n}\varpi_{N+k+1}y_{N+k+1} \\ \varpi_{N+k+1}y_{N+k+1} \end{pmatrix}$$

It should be noted, that the matrix, which is inverted (Q^{k+1}) has much smaller dimension $((n+1)*(n+1))$ in respect to the dimension of x_e and y becuase $n<<N+k$. The dimension of the other up-dated matrix S^{k+1} is even lower $((n+1)*1)$. The bulk of the information is passed in compressed form trough matrices S^k and Q^k from the previous step's calculation speeding up the algorithm execution.

The up-dated matrix (F^{N+k+1}) has dimensions $((n+1)*(N+k+1)$. The bulk of the information is passed in compressed form trough the **recursive** up-date from the previous step's calculation. This speeds up the algorithm execution. One of the dimensions of the matrix F, however, is normally large because it represents the influence all data collected at the moment k in *on-line* (x_j; $j=1,2,\ldots,N+k+1$) has to the new rule. Practically, this problem could be handled by the mowing time window.

In this way, the whole rule-base is preserved with substitution or an up-grade with one rule only and this gradual change takes place only when informative data are collected. In practice, this takes place very rarely - for less than 5% of the new data as seen from the examples shown in the next part of the book.

It should also be noted that the up-grade of the rule base with new rules is not tending to an overload, because the closeness to already existing centres is a hard restriction. In this way, a new rule appears, only if significant for the object new area in the data space has place and it has not been covered before. This could be due to the change (not incidental, but supported by enough new data) in the objects behaviour due to wearing, fault, ageing etc.

The minimal number of training data point/samples should be larger than the number of model inputs (n) in order to avoid computational problems with inverting the matrix Q (7.8):

$$N>n \tag{7.14}$$

7.6 FRB Model Up-grade; *'Learning Trough Experience'*

Condition (7.14) determines the lowest limit of the number of data points/samples necessary to start the *on-line* training and up-grade of the rules' structure. In order to test the ability of the algorithm to up-grade the rule-base starting with a low number of initial *flexible* rules two sets of experiments have been carried out:

- ✓ RH: Relatively High initial number of training data points/samples (N>>n; N ~300÷500);
- ✓ VL: Very Low initial number of training data points/samples (N ≈ n; N > n; N~10÷20).

The initial rules' structure could be formed on the basis of expert knowledge (if such one exist in appropriate form) or could be extracted from experimental data by *off-line* data-driven approach (see Chapters 4 and 5). It could be also a combination of both in a hybrid scheme (Angelov and Guthke, 1997).

In the Chapter 8 a sub-system for indoor climate control (ICC) and a dynamical time-series model are considered with these experimental set-ups (RH and VL). The results illustrate that the model structure (*flexible* rules and linguistic terms) is up-graded successfully starting with both Relatively High and Very Low initial number training data points/samples.

This property of ***e*R** models makes them very appropriate for fields like behaviour-based modelling, robotics, agents etc. This is also a key feature, which makes them an effective tool for building *intelligent* and *smart* adaptive systems. It gives real sense to the commonly used, especially in neural-network-related literature, term *learning trough experience.*

7.7 Rule Structure and Parameters Tuning and Refinement

It is desirable, especially in fault detection and diagnostics, to have a transparent model which is as simple as possible while maintaining a pre-defined level of precision. It is also computationally more efficient to manipulate smaller model structure (having lower memory requirements and number of operations), suppose it gives the same or quite similar performance.

Two mechanisms for model refinement will be considered here. The first one is *non-iterative*, while the second one requires iterations and therefore is optional.

7.7.1 Similarity-based Simplification of Linguistic Terms

The number and structure of the *flexible* rules are supposed to be determined by the data space clustering and are fixed during these refinements. The number of linguistic terms of every *flexible* variable, however, could be different (smaller than that determined initially by clustering), while the overall performance of the model could not suffer significantly or at all. This fact has been considered already for the *off-line* algorithm (Section 5.4.1).

In order to maximise the transparency, which also leads to minimising the memory cost, it is necessary to minimise the number of membership functions describing each linguistic variable. If suppose the same type of Gaussian membership functions and equal fixed spreads is used (which is an often-used assumption) then the number of linguistic terms and the structure of *flexible* rules could be simplified based on the closeness of the rules centres only.

For the more general case of different spreads or different types of membership functions, it could be done using the degree of similarity between each membership function (3.15). The fact that this could be done ***non-iteratively***, makes possible to include this procedure into the *on-line* algorithm (Fig. 7.9).

After a new centre has been determined in *on-line* mode by up-date of potentials, the similarity of the membership functions describing the respective *flexible* rule and the membership functions based on already existing centres is calculated. If the membership function of new linguistic term is similar to one that already exists it is ignored and the existing FLT is used. If no similar membership function exists, the rule innovation mechanism is applied (Section 7.4) and this rule is added to the model.

The threshold for similarity is established based on the distance between two centres. It is recommended to use 10 to 15% of the whole range of possible values of a variable. It should be noted that the threshold over the closeness as a simplification criterion should not be too high (more than 25%). Otherwise, the model precision could suffer. Conversely, it should not be too low (less than 5-10%), as there will not be sufficient simplification of the model.

Generally speaking, it is a balance between transparency and the precision. It is interesting to note that significant reduction of the number of linguistic terms used and respectively simplification of the model structure could be done without practically sacrificing the model precision (Angelov and Buswell, 2001b). Models structure simplification increases its generalisation and descriptive potential, while precision itself very often leads to over-fitting and is significantly more data-dependent and bounded to the specific problem.

The number of rules influences the precision of the model and is determined by the potential of data samples. The model simplification process seeks to minimise at the next phase the number of membership functions associated with each input variable separately. The model structure could be further simplified and transparency of the FRB model could be improved by simplifying linguistic expressions using so called *'don't care'* representation for specific linguistic variables.

An application of this procedure to ICC components modelling is considered in the next Chapter 8.

7.7.2 Parameters Refinement (Tuning) by Non-linear Optimisation

Similarly to the *off-line* case, a refinement or tuning of the membership functions shape and parameters could (optionally) be done by a non-linear optimisation technique, such as (real-coded) GA, gradient-based *error back-propagation*, simulated annealing or any other search algorithm. Real-coded GA is preferred than gradient-based approaches mainly because it successfully avoids local extremums, which in identification problem could be associated with incorrect model structure.

In this procedure centres of membership functions as described above are used as initial estimation only. They are further optimised in a close range around these values. This, normally, leads to improved precision of the model (up to 2-3 times) both in training and validation, because of the ability of GA to find global optimums or to be close to them.

This effect is, however, for the expense of computational efforts (because it is an **iterative** search algorithm) and decreased linguistic transparency. Therefore, it is recommended to perform this tuning much rarely in comparison to the time-step of the collection of data samples ($K>>1$, where K is a problem-dependent parameter). K takes into account the time constant of the process, time consumed by the GA and importance of the fine-tuning. For example, for classification and decision support problems, by differ from control ones, the time could be not as important as the precision. Then it could be justified to make this refinement much more often.

If a real-coded GA (Michalewicz, 1996) is used, its chromosomes could consist of input and output parameters:

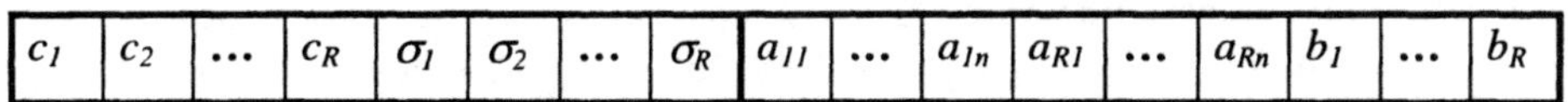

Fig.7.8. Chromosome with encoded FRB model parameters (both antecedent part - left and consequent one - right)

where $c_{1,...,}c_R$ denote centres of clusters used as a basis in formation of rules;

$\sigma_1,\ldots,\sigma_R$ denote spreads of membership functions;
$a_{11},\ldots, a_{Rn},\ldots,b_1,\ldots, b_R$ denote parameters of consequent part of TSK model.

Centre-of-gravity-based recombination operator (see Section 3, Chapter 4) could also be used for speeding-up its convergence.

7.8 Flow-chart of the Algorithm

The proposed approach is performed in 8 phases represented in the Section 7.2. Flow-chart of this algorithm is depicted in the Fig. 7.9. Respective phases of the *on-line* procedure are given as numbers in circles. The steps, which form *off-line* identification procedure, are surrounded in a separate block.

All steps are ***non-iterative*** except fine-tuning by GA (generally, it could be any numerical search technique), which, however, could be performed for each K steps. r is a positive integer constant ensuring the repetition of the process after each K steps (recommended value for r is 100, but it highly depends on the problem specifics, computer platform etc.).

First N data samples are supplied *off-line*. They could be just some tens (if VL strategy is used) or some hundreds or thousands (if RH strategy is applied). In the first case the model will be build '*on fly*' by *learning trough experience*. In the second case an up-date, *on-line* performance monitoring and analysis could be achieved.

The rest k samples are provided *on-line*. Due to the ***non-iterative*** nature of the algorithm the computational demands are extremely low: the computation time required for performing them on a standard PC (550 MHz; 128Mb RAM) is less than 1 s. This makes the algorithm applicable to a wide range of industrial processes, which have faster dynamics.

The *on-line* calculation starts after the respective new data point/sample is collected, filtered and passed. First, the potential of this new data point is calculated using (7.1a)-(7.1b). Next, the potentials of already determined centres of membership functions of *flexible* rules (in fact, clusters in the data space) have been up-dated (using (7.2)-(7.3)) taking into accounts the influence the new data point impact on them.

The most important and distinctive part of the proposed approach is the rule-base ***innovation*** by adding a new rule (rule-base up-grade applying a real *learning trough experience*) and/or replacing less effective rules by more-descriptive ones (rules ***modification***).

Application of similarity-based model reduction improves the transparency of the model, without significant degradation in the model precision. Additional benefits of this simplification are the reduction in execution time and memory requirements.

Finally, the resulting up-dated (evolving) *flexible* model is exported for use.

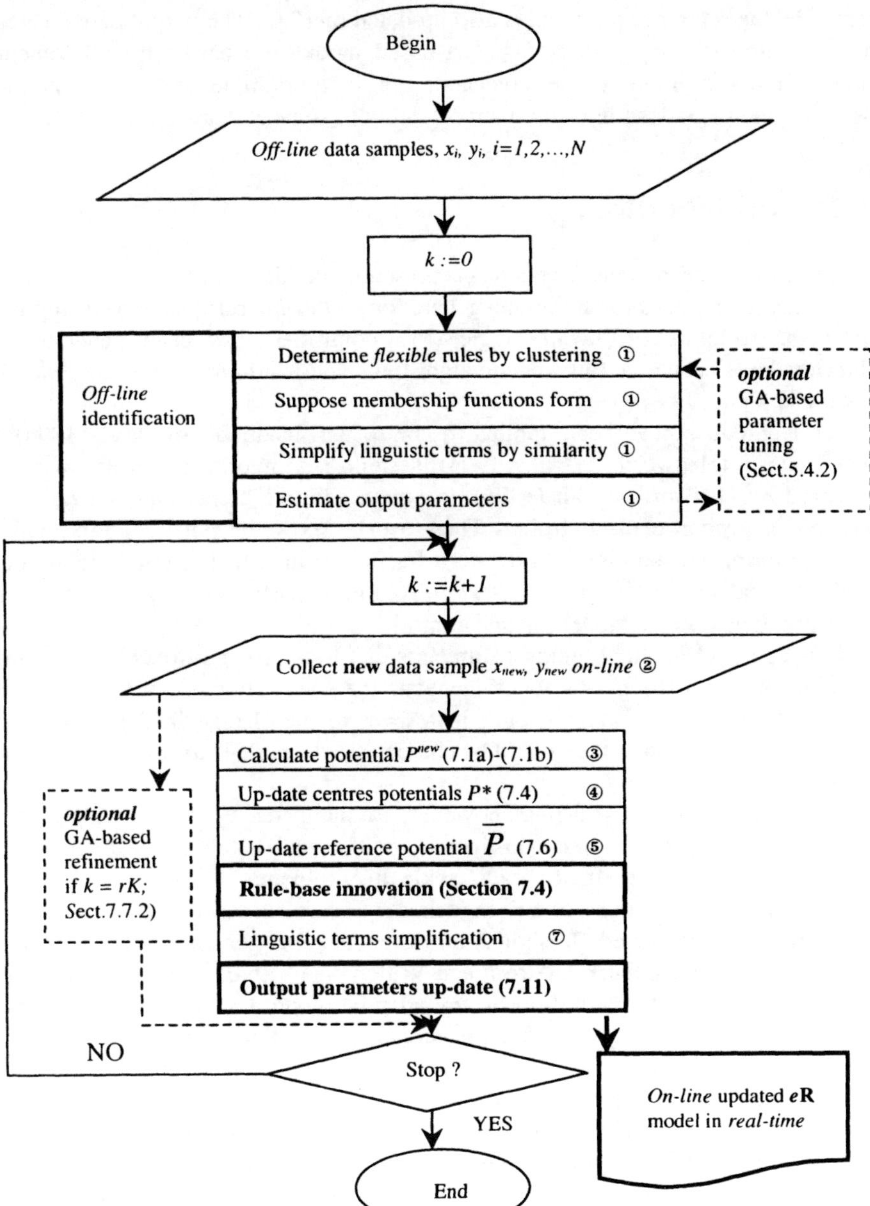

Fig. 7.9. Flow-chart of the *on-line* identification procedure

Meanwhile, the reference potential is also up-dated by (7.6). The output parameters are recursively up-dated by (7.8)-(7.11). GA-based model and parameter refinement is optional. It could improve the precision, but is more time and computationally consuming and degrades linguistic interpretability to some extent.

7.9 *e*R Control Algorithm

An application of ***e*R** models for control is presented in this section.

Recently, techniques for automatic generation of *flexible* rules have been applied to controllers. Structure of *flexible* logic (FL) controllers has been generated and parameters have been tuned by genetic algorithms and neural networks (Kosko, 1992; Cooper and Vidal, 1996).

Such techniques have been applied to control air-handling (So et. al., 1994) and heater battery (Hepworth et. al., 1994) systems. A hybrid controller has been considered by Hepworth et. al. (1994), in which additive corrections are made to a *conventional* proportional controller. These works show the potential for improved control performance and highlight the need for further investigation into the use of FL in order to find more efficient and *flexible* solutions, particularly to the problems of air-conditioning systems modelling and control.

These approaches allow tuning parameters of FL controllers to achieve the desired quality of control; it is also possible to generate the whole structure of the *flexible* rules of the controller and to adapt it later in respect to the change in the environmental parameters. In this context they could be characterised as self-learning, *evolving* control schemes.

On-line application in *real-time*, however, are hampered by the high computational costs of genetic algorithms and non-linearity of the models. ***e*R** models considered in this chapter could be used effectively to tackle this problem.

One possible application of ***e*R** models for control based on an indirect control scheme will be considered. It should be noted that, in general, there exist various schemes for adaptive control (Astrom and Wittenmark, 1990). Also it is important to mention that a practical realisation of any adaptive control scheme suppose a careful investigation of the stability, which, however is highly problem-dependent. This control scheme for example supposes minimal phase system: such one, which has stable zeros in the transfer function (Astrom and Witenmark, 1990).

Indirect control (Andersen et. al., 1994) supposes model-free concept. The structure of the controller, however, remain to be determined (Fig. 7.10):

If *flexible* logic controller (FLC) is considered then its structure (rules, linguistic terms, membership functions etc.) have to be determined. Using ***e*R** model it is possible to generate it *on-line* in parallel with the control itself.

First few (normally, 3-5) initial time steps forms the input-output data for ***e*R** model training. During this initial few time steps it is possible to apply any conventional control algorithm, like P, PI or a FL control with some suggested structure and parameters, which will be later updated.

This scheme is a real *'learning trough experience'* and is illustrated in the Fig. 7.10.

The object of control (air-conditioning plant, (bio)process etc.) have to be controlled such that to keep the set point value y^{sp} by a control signal u_k. Disturbances (F_k) could be measured or not.

The idea of indirect learning using ***e*R**-based *flexible* controller is simple:

After few initial steps (during which any conventional – P, PI, some FL etc.- control algorithm is used) a set of training (input-output) data sets has been collected. They comprise of inputs (y_k and y^{sp} and possibly F_k, if available and measured) and the output – desired control signal (u_k^d) as shown in the Fig. 7.10.

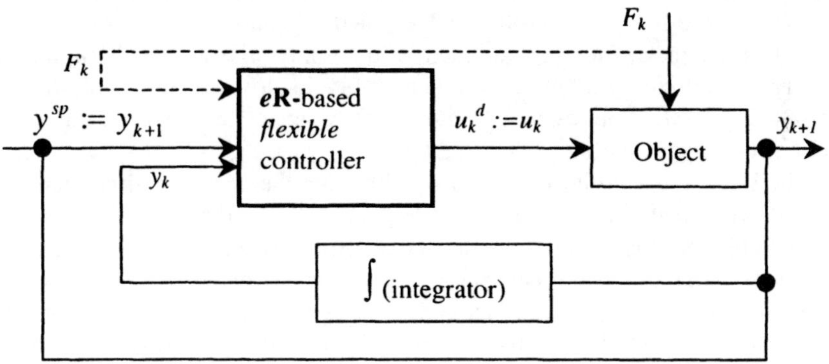

Fig. 7.10. **eR** control scheme based on indirect learning principle

The logic behind the indirect adaptive control is as follows (Andersen et al, 1994):

Having a sequence of real outputs of the object of control (y_k) and control signals (u_k) and possibly measured disturbances (F_k), which form, in fact, the input(s) to the object, it could be supposed that

Since after applying at the time instance k certain control signal u_k (under disturbances F_k and having output y_k) the output has been at the next time instance y_{k+1}

Then in order to achieve the same output $y_i = y_{k+1}$ (under the same disturbances F_k and output y_k) it is necessary to apply the same input (control signal u_k).

Therefore, the training samples are formed as:

⇒ **inputs**:

- y_k;
- $y^{sp} := y_{k+1}$;
- F_k (if available);

  i.e. the set-point is established equal to the output from the process at the next time instance;

⇒ **output**: $u_k^d := u_k$;

i.e. the desired control signal is set to the real control signal at the time instance k.

It is proposed to use ***e*R** model and to generate it in *on-line* mode producing as an output the control signal to the object of control. Thus, the ***e*R** model in this scheme works as a *flexible* logic controller and is called ***e*R** controller.

In this scheme *on-line* measurement of the following variables is supposed: control signal really applied (u), output (y) and possibly disturbances (F_k). The structure of the ***e*R**-based *flexible* controller is automatically generated and up-dated using the approach presented earlier in this chapter.

The role of the initial control algorithm (for the first few time steps) is just to generate several starting training samples (n=2 or n=3 if disturbance is also considered and therefore N>3 is required to start the algorithm according to (7.14)). It is replaced after these initial 3-5 steps. Therefore, it is not critical which algorithm exactly will be used. The ***e*R**-based approach has high adaptive ability and corrects later the initial rules, when necessary based on experimental data collected during the process run.

This approach has been tested with an illustrative example of a heat transfer into a single zone unoccupied room.

Example 7.1 (based on (Loveday and Virk, 1992))

A simulation of a test cell representing a single zone unoccupied room is considered in the form of the following regression model:

$$T_{k+1} = 2.03T_k - 1.25T_{k-1} + 0.22T_{k-2} + 0.74u_k - 0.34u_{k-1} - 0.33u_{k-2} - 0.04 \qquad (7.15)$$

where T denotes room temperature;
u is the control signal (coil capacity)

This simulation model is considered instead of the real test cell as in (Loveday and Virk, 1992). It is required to keep constant temperature in the room by external heath transfer. Initial temperature is 30 ^oC. The following set-points are considered (they provide a thermal comfort for potential occupants):

- $T_1^{sp} = 24^0 C$;
- $T_2^{sp} = 22^0 C$;
- $T_3^{sp} = 23^0 C$;
- $T_4^{sp} = 20^0 C$

A standard proportional controller is considered for the first three steps ($k_P = 0.5$). From the time-instance $k=4$ onwards an *on-line* training procedure starts, which has been described in Section 7.2. The training data are triplets $\{T_k, T_{k+1}, u_k\}$.

The algorithm (Fig. 7.9) generates centres of *flexible* rules and forms membership functions (Gaussian type with fixed spreads have been used for this example for the sake of simplicity) and estimates (non-iteratively) parameters of the output. Then at each time-step it up-dates the structure and parameters of this FRB model recursively taking new triplets for $k=5,6,...$

This FRB model with a structure and parameters constantly up-dated is used to produce the desired control signal $u_k^{\cdot\cdot}$, which is applied to the object (in this case to the model simulating the unoccupied room).

The profile of the control signal (coil capacity in kW) which is applied is shown in the Fig.7.11.

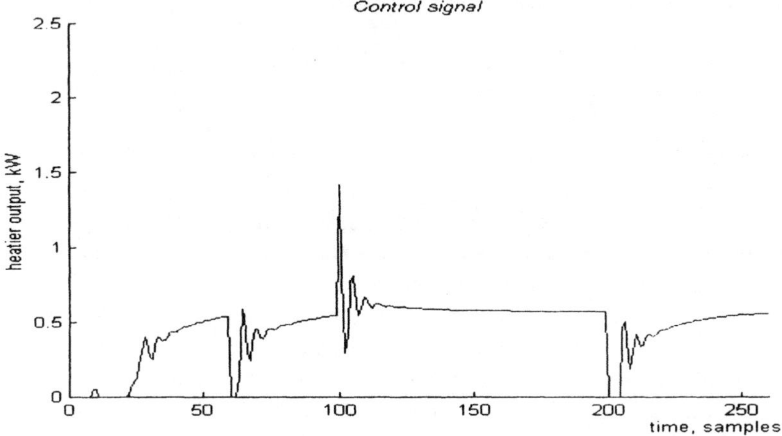

Fig. 7.11. Energy consumed by the coil: control action of the **eR** controller

Under the assumption, that the thermal characteristics of the room are represented adequately by the simulation model (7.15), this control action has to provide the following temperature inside the room (Fig. 7.12):

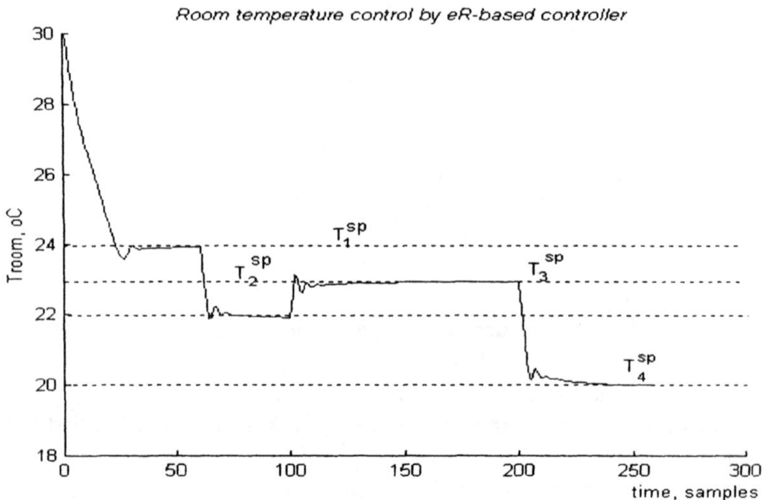

Fig. 7.12. Temperature of the room under **eR** control.

7.10 Conclusion

The approach for rule-base evolution by **recursive** *on-line* adaptation of rule structure and parameters and simultaneous model simplification is presented in this chapter. It is central in the concept for ***e*R** models presented in this book.

By this approach a transparent, compact and accurate model is designed autonomously by evolution based on the experimental data and permanent *innovation* of the rule base together with the membership functions' parameters. It could be used as a computational basis for development of truly (or at least more) *intelligent* adaptive systems in combination with respective sensors and filters for *on-line* data collection.

It relies on the specific of TSK type of *flexible* rule-based models to have a dual nature and benefits from the possibility to consider separately structure identification (by data space clustering) from parameter estimation (done by linear least squares) and to use **non-iterative** (in general) procedures for both of them.

On-line adaptation of the rule structure by its up-grading and by replacement (rule-base modification) of a rule has been illustrated with an engineering example from ICC systems. A more thorough presentation of *on-line* modelling different components of indoor climate control systems is presented in the next chapter.

One possible application of the *on-line* modelling approach is for control purposes. An adaptive control scheme based on the indirect learning approach and ***e*R** models (called ***e*R** control) is discussed and illustrated with a simple example of an unoccupied single zone test cell.

PART III ENGINEERING APPLICATIONS

In this part of the book the results and discussion of engineering applications of the proposed methodology for *on-line* identification of FRB models (called briefly ***e*R** modelling) are presented. Non-linear *off-line* approach has also been illustrated.

In Chapter 7 the concept of ***e*R** models have been tested and validated with data from building services engineering sector: modelling the duty and outlet air temperature from a heating/cooling coil (an often-used component of indoor climate control systems). *Flexible* rule-based models of the other basic components of such systems (fans, boilers, and fan-coil sub-system with valve control) are considered next.

Model structure simplification (reduction of linguistic terms) has been illustrated with *real* data from a fan-coil sub-system. Dynamic regression model of cooling coil control signal is presented next. The two strategies (RH and VL) which concern the number of data samples (Chapter 6) have been compared. The VL strategy illustrates the ability of ***e*R** models for a true *'learning trough experience'*.

Centre-of-gravity based crossover operator has been tested with a number of numerical functions and with a real problem of optimal scheduling of a hollow core ventilated slab system.

A system theory approach to the problem of indoor climate control has been presented at the end of this chapter. ***e*R** models are seen as a promising tool for design of *intelligent* ICC (***i***-ICC) systems.

In Chapter 8 application of ***e***volving **R**ule-based models to fermentation processes has been demonstrated. A *real-time* up-date of a *flexible* rule-based model (designed initially in *off-line* mode) of a fermentation process using data samples collected in *on-line* mode has been presented next.

Application of FRB and ***e*R** models to development of *intelligent* risk assessment systems has been treated in Chapter 9 on the examples of banking and civil aviation. The notion about evolving *intelligent* Decision Support Systems for tendering evaluation in large-scale construction project management has been also considered.

8 MODELLING INDOOR CLIMATE CONTROL SYSTEMS

Indoor Climate Control (ICC) systems are major energy consumers. According to the latest data published by the International Energy Agency (http://www.iea.org) globally about a half of the primary energy is used in buildings. Therefore, it is of vital importance to understand, model and control the performance of such systems effectively.

An example of a simple single duct heating, ventilating and air-conditioning (HVAC) system, which is an important part of ICC systems, is shown in the Fig. 8.1.

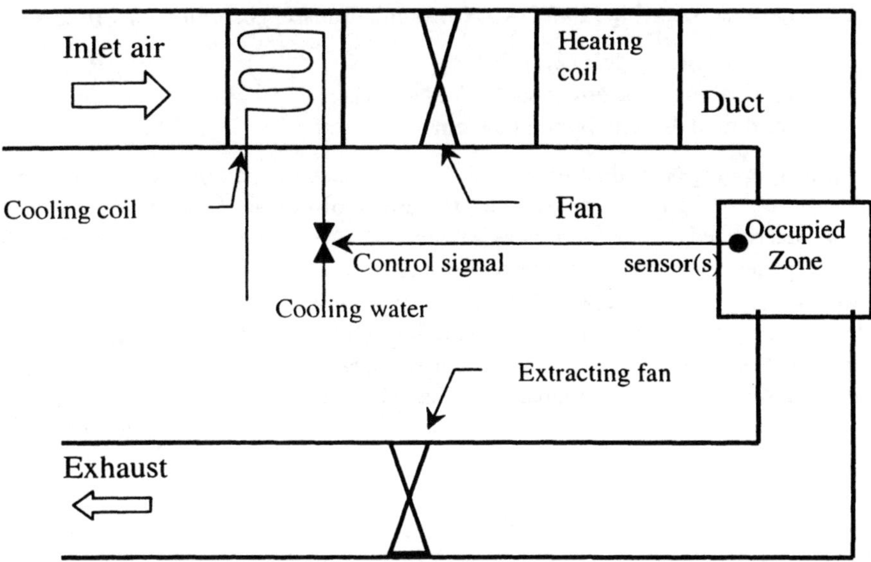

Fig.8.1. A single duct air-conditioning system

Modelling and simulation is an important element of the design and performance analysis of HVAC systems as well as a part of the overall building design and has been the subject of much research in last two decades (Hanby and Wright, 1988; Brandemuehl et. al., 1998). The emergence of new methods for modelling such as *fuzzy* models offers greater possibilities for a more *flexible* and generic description of HVAC system components (Angelov et. al., 2000a; Ghiaus, 2001).

Simulations of HVAC are usually performed in combination with building simulations. This leads to considerable computational demands, because these are often used for energy predictions over annual operational cycles. *Flexible* rule-based models have the potential to reduce the computational demand of the simulation by mapping the inputs and outputs of the system components directly.

By differ from 'pure' *black-box* models (including neural networks), however, they give an insight to the process relations and are easily understandable. Additionally, ***e***volving **R**ule-based (***e*R**) models as well as non-linear identification using GA combined with an appropriate encoding do not require the model structure to be known and fixed *a priori* as it is the case with the *black-box* models.

8.1 Modelling Components of HVAC Systems

Component models used in HVAC system simulation are, generally, classified as:

- ✓ first principle models (2.1)-(2.4);
- ✓ empirical or *black-box* models (2.11)-(2.12);
- ✓ hybrid models, which include elements of both.

First principle models realise mass and energy balance equations of air and water as well as some of their properties like pressure, moisture content etc. The operating point is determined as the numerical solution of a set of equations, which is a time-consuming and, sometimes, imprecise procedure.

In reality, processes associated with some typical components, such as coils, boilers or fans, can be too complex to be described by analytical methods. Polynomial representations (Wright, 1991), and, recently, neural networks (Diaz et. al., 1998) and *flexible* rule-based models (Ghiaus, 2001) have been employed alternatively.

Empirical models are frequently used when the underlying principles are not well characterised, or when the computational effort required implementing first principles models is prohibitive. For example, pressure and power as functions of the flow rate in fans, refrigeration effect as a function of evaporating and condensing temperatures in compressors, efficiency of boilers as a function of their maximal load and water temperature etc.

Limitations of this type of models (also referred to as *black-box* models) include their lack of transparency as well as limited range of validity. Parameters of the NN and polynomial models have no physical meaning, while parameters of first principle models as well as *flexible* rules have a clear interpretation.

FRB models are based on the rules, which represent the dependence of the model variables in plain English. One possible way of implementing FRB models is the accumulation and the expression of existing knowledge. In this case their parameters are tuned based on experimental data, but the rules are formulated by experts.

Formalisation of expert knowledge is often a subjective and ambiguous process. Therefore, the alternative data-driven approach is founding wider application. This approach allows a more *flexible* and objective way of model structure determination based on the data only.

The new approach for *on-line* construction of rule-based models, presented in the Chapter 7 is an alternative to both first principle and polynomial models usually used in component modelling of HVAC systems. The resulting models are transparent and existing expert knowledge could easily be incorporated into or extracted from the model.

Application of this type of FRB models generated by the non-linear *off-line* identification approach to model the basic components of HVAC systems (coils, fans, boilers) is presented and discussed in this section.

Models originally developed for ICC component simulation have been applied to *on-line* condition monitoring. Condition monitoring uses a reference model, whose output is compared to the monitored process (Salisbury and Diamond, 2001). Large differences between these variables can indicate non-optimally operating equipment and be used to generate an alarm.

Data from the target system are normally needed to tune the models in order to predict the performance to an acceptable degree of accuracy. It is an advantage of the ***eR*** models that they can make use of data from normal operating conditions, negating the requirement for special testing in addition to the advantages over *black-box* models discussed earlier.

8.1.1 Heating/cooling Coil Modelling

On-line identification is demonstrated on *real* data from an air-conditioning unit (Angelov et. al., 2000b). Heating and cooling coils (Fig. 2.1) are often used component of HVAC systems. Cooling and dehumidification of the air approaching the coil are important processes in terms of the comfort of the occupants of the air-conditioned space. These processes can also be energy intensive and hence it is often of interest in building energy and performance simulations.

Under normal operation, the outlet from the coil air temperature (which is supplied to the Zone) is controlled to some pre-determined set point by regulating the mass flow rate of chilled water through the coil. This is achieved via a control signal that commands an electrically driven actuator, which moves the valve stem. Internally, the plug at the end of the valve stem moves and diverts the water flow from one port to the other. The supply air temperature should remain fairly constant, while the inputs will vary and the differences will manifest in the value of the control signal.

Conventional model consists of four non-linear algebraic equations describing the mass and energy balance in a steady state (2.1)-(2.4). The procedure of iterative solution of algebraic equations, however, is time consuming and not precise. Additionally, the non-linear models of the value, which often includes a hysteresis and psychometrical dependencies, should be taken into account

Alternatively, the output variables could be calculated straightforward using ***e*R** models in *on-line* mode.

Measurements of *real* data for the air and water inlet temperatures (T_w^{in}, T_w^{in}), volumetric flow rate of air (m_a) and the control signal to the control valve (U_{cc}) are used as inputs to the model (courtesy of ASHRAE for the use of data, generated from the ASHRAE funded research project RP1020). The model is then used to predict outlet from the coil air temperature (T^{out}). Samples were been taken at one-minute intervals for two 24-hour periods, a few days apart. Some assumptions have been made in order to simplify the problem (Angelov et. al., 2000b):

- ✓ The model is based on the static data alone and so the data that exhibits dynamic characteristics have been filtered;
- ✓ In principle, the coil performance is affected by the air moisture content. When the conditions allow, condensation forms on the surface of the coil, reducing the moisture content in the supply air. This additional latent heat exchange reduces the sensible heat exchange and therefore affects the air temperature also. This effect is not considered in this example;
- ✓ In the test system, the fan speed was also under a control regime that allows the fan speed to vary. The heat surrendered to air is proportional to the work done by the fan on the air. However, under the control administered, normal operating conditions, over typical office hours, for the same seasonal period, the variation in the temperature across the fan will be at a minimum.

One complication for the modelling of the system is the highly non-linear characteristics. The cooing coil gives an exponential response to an increase in water mass flow. The control valve and actuator are designed to counter these effects leaving a linear water mass flow/heat output response. Practically, these criteria are seldom met and highly non-linear characteristics, often with some 'dead-bands' at either end of operation are typical.

8.1.1.1 Modelling Outlet (from the Coil) Air Temperature

The same data for the ambient temperature (a *typical* summer day of 3 August 1998) as in the Example 5.1 are considered. Constant water inlet temperature (T_w^{in}) of 8^oC and a dry inlet air ($g=0$) are supposed. Outlet from the coil air temperature (T^{out}), which is normally supplied to the zone (Fig. 2.1), is modelled based on a *real* control signal (U_{cc}) aiming to bring the outlet temperature supplied to the zone down.

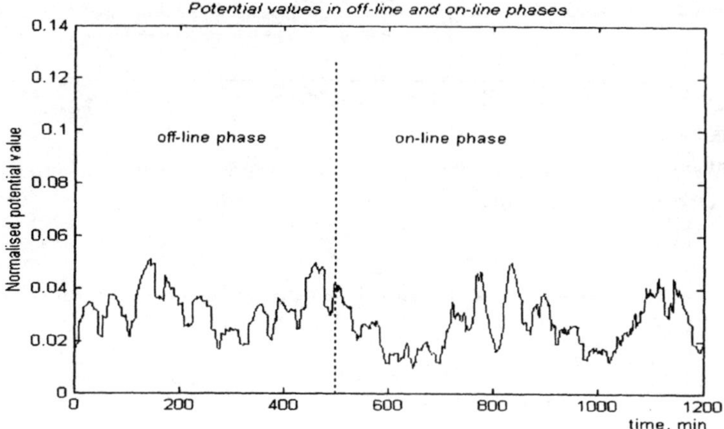

Fig 8.2. Potential values of *off-line* and *on-line* collected data samples

The **eR** model is considered having the following inputs:

- inlet air temperature (T_a^{in});
- air flow rate (m_a);
- control signal (U_{cc}) to the valve regulating the water mass flow rate.

A set of $N=500$ data samples have been collected in *off-line* mode and used for training the FRB model. For the next $k=0,1,...700$ minutes the performance of this FRB model have been compared to the performance of an evolving (**eR**) model and to the original data. Normalised potentials of the experimental data are given in the Fig.

8.2. The output (T^{out}) generated by both models have been plotted versus the original data in the Fig. 8.3.

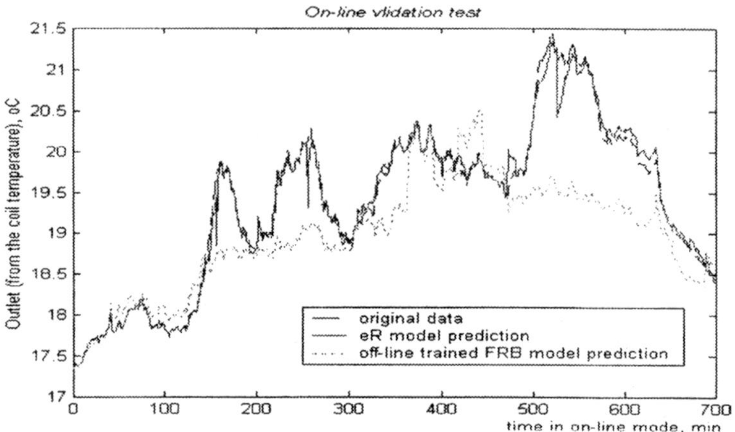

Fig 8.3. Outlet temperature: **eR** model (solid line); *off-line* trained FRB model (dotted line) and the original data (dashed line)

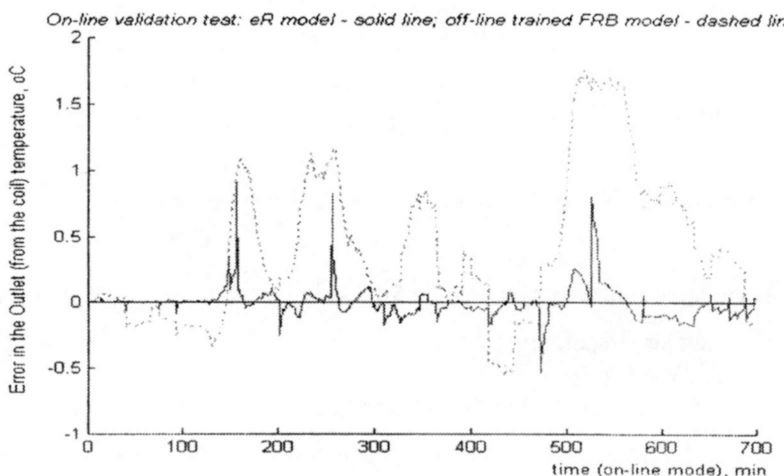

Fig. 8.4. *On-line* validation test: eR model-solid line; *off-line* trained FRB model-dashed line

It could be seen that ***e*R** model is significantly better (Fig 8.4 shows the absolute error in prediction), because it is able to adapt, to learn in the process of data acquisition and renovate its structure (Fig. 8.5) and its parameters. The mean root squares error (MRSE) of the ***e*R** model in this validation test for the outlet from the coil air temperature is *0.004435 K* compared to *0.026111 K* for the *off-line* model. The improvement is almost 6 times (589%)!

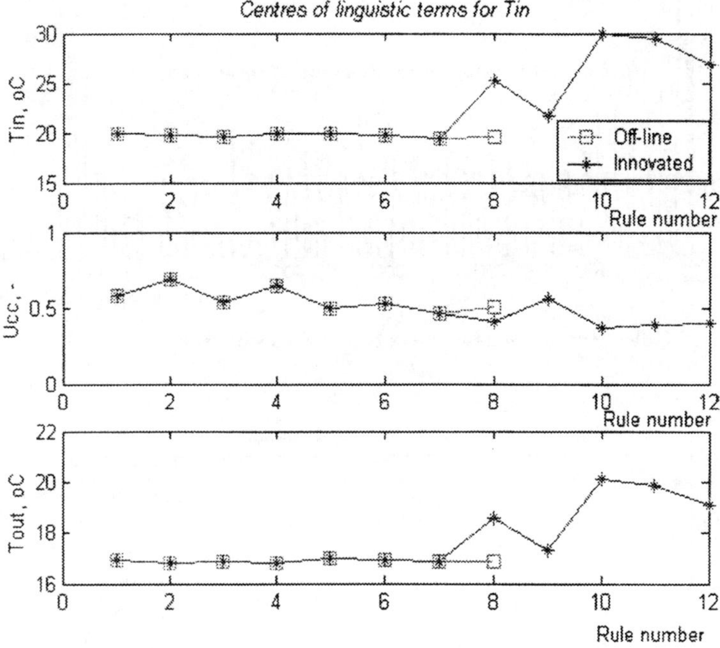

Fig. 8.5. Rules *innovation* represented for each linguistic term

The rules' innovation process based on the data points' potential is illustrated in the Fig. 8.6. It is seen that the first new data sample is informative enough (it has potential higher than $\overline{\varepsilon P}$; threshold value $\bar{\varepsilon}$ =50% is chosen as in (Chiu, 1994)) as well as the potential of data points around the minutes 600-660 from the start of the *on-line* mode. One other new data sample with high potential (lying between the dashed lines) and not too close to previous centres have also been chosen to replace the respective rules (Fig. 8.7).

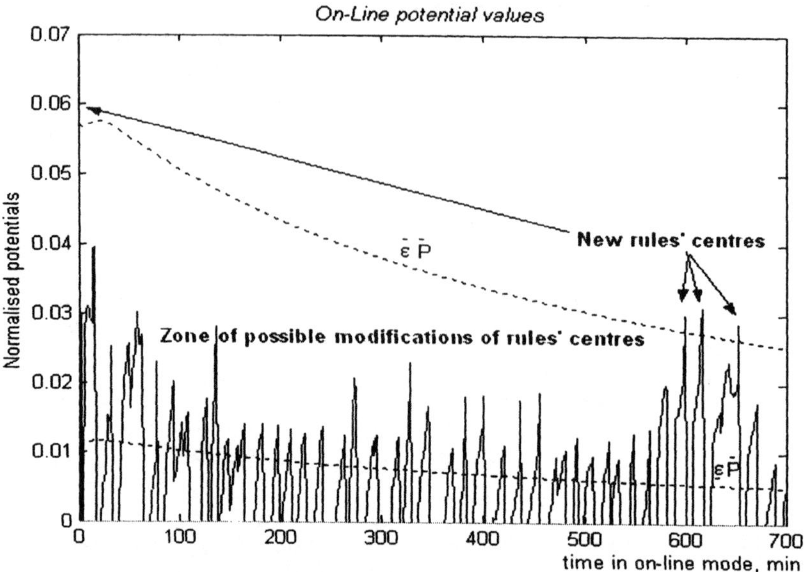

Fig. 8.6. Rules *innovation* based on potential values of centres

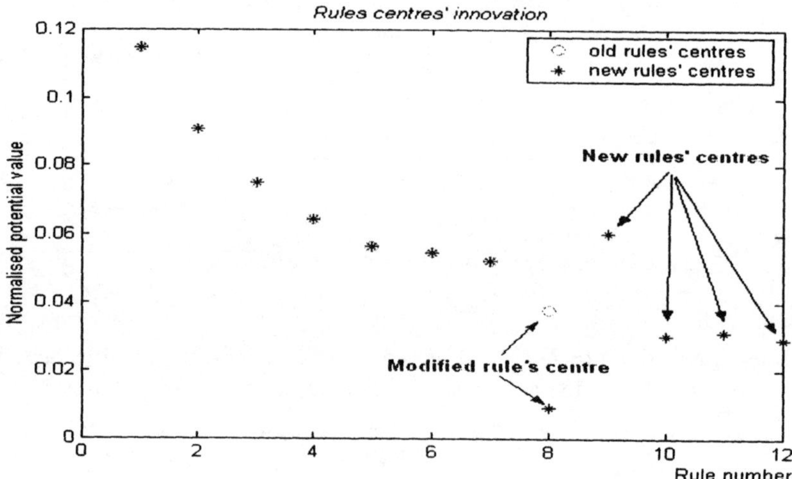

Fig. 8.7. Rules *innovation* based on the cooling coil FRB model

8.1.1.2
Modelling the Heat Transfer in a Heating/cooling Coil

A set of 8 coils with a face area $A=0.5\ m^2$ and inlet air pressure $p^{in}=1000\ N/m^2$ has been considered as an example (Angelov et. al, 2000b). A set of 79 steady state working points has been used as experimental data.

The total heat transfer rate (Q) could be modelled as a function of air and water inlet temperatures (T_a^{in}, T_w^{in}), mass flow rates (m_a, m_w) and moisture content in the inlet air (g) (Hanby and Wright, 1989).

In this example there are 6 linguistic variables: 5 inputs ($T_a^{in}, T_w^{in}, m_a, m_w$, and g) and one output (Q). It is supposed that each *flexible* variable has 9 linguistic terms ($n=5$, $m_i=9$; $i=1,2,\ldots,5$) *which are tabulated* in Table 8.1.

The non-linear *off-line* identification of a FRB model of not more than $R=30$ rules is considered using GA and the approach presented in the Section 4, Chapter 4. As a result of parameter identification it eventually appears that the actual number of rules is smaller than *30* ($r<R$) due to the coincidence of some parameters or rule indices. This adds to the *flexibility* of the model structure definition. The output (Q) is represented by triangular membership functions.

Table 8.1. Linguistic labels of *flexible* variables (m_a, m_w, T_a^{in} T_w^{in}, g, and Q)

Flow rates	Inlet temperatures	Moisture content	Heat transfer
Extremely Small	*Very Hot*	*Extremely Dry*	*Extremely Low*
Very Small	*Hot*	*Very Dry*	*Very Low*
Small	*Warm*	*Dry*	*Low*
Rather Small	*Rather Warm*	*Rather Dry*	*Rather Low*
Medium	*Neutral*	*Neutral*	*Medium*
Rather Big	*Rather Cold*	*Rather Wet*	*Rather High*
Big	*Cold*	*Wet*	*High*
Very Big	*Very Cold*	*Very Wet*	*Very High*
Extremely Big	*Chilly*	*Extremely Wet*	*Extremely High*

An initial population (consisting of 30 chromosomes) is generated randomly. Each chromosome consists of 60 genes (30 for the rules and another 30 for membership functions centres; spreads are fixed to $\sigma_{lj}=0.1(\bar{x}_l - \underline{x}_l)$, $l=1,2,...,6;\ j=1,2,...,9$). Their evolution is driven by a selection of the fittest and is performed by recombination and mutation.

Finally, a set of *flexible* rules is generated which describes the data samples with product moment correlation *0.9894*. Several runs of optimisation have been carried out and correlation coefficient ranges from *0.9515* to *0.9894*.

The chromosome with the best fitness value represents *flexible* rules and centres of their membership functions of the input variables. The second rule, for example, is:

$$R_2\text{: } \textbf{\textit{IF}}\ (T_a^{in} \text{ is Very Hot})\ \textbf{\textit{AND}}\ (g^{in} \text{ is Rather Wet})\ \textbf{\textit{AND}}\ (m_a^{in} \text{ is Medium})$$

$$\textbf{\textit{AND}}\ (T_w^{in} \text{ is Cold})\ \textbf{\textit{AND}}\ (m_w^{in} \text{ is High})\qquad \textbf{\textit{THEN}}\ (Q \text{ is Very High}) \tag{8.1}$$

The centres of membership functions defining *flexible* sets of the airflow rate, for example, are given by *{0.899; 1.269; 1.445; 1.845; 2.525; 2.887; 3.611; 4.056; 4.492}* kg/s. Parameters of triangular membership functions of the output variable Q are chosen to cover symmetrically the expected output range of Q: *{5; 12.5; 20; 27.5; 35; 42.5; 50; 57.5; 65}*, kW. They are fixed (excluded from optimisation).

This Mamdani type FRB model represents the main dependencies between input linguistic variables and the output, namely the proportional dependencies between Q and T_a^{in} and exponential one between Q and m_a^{in}. Saturation exists at high levels of m_a^{in}. It occurs earlier when higher moisture content in the air leads to condensation. This is consistent with the expected behaviour of a cooling coil (Hanby and Wright, 1989).

8.1.2 Ducted Fan Modelling

Ducted fans (Fig. 8.8) are considered as another illustration. The conventional model consists of two non-linear algebraic equations describing the energy balance in a steady state (Wright, 1991):

$$p_{tf} = p^{out} - p^{in} \tag{8.2}$$

$$q = mC_p(T^{out} - T^{in}) \tag{8.3}$$

where p_{tf} denotes fan total pressure;

q denotes fan absorbed power;

C_p is the specific heat;

m, T^{in} and T^{out} are air mass flow rate and temperature as above.

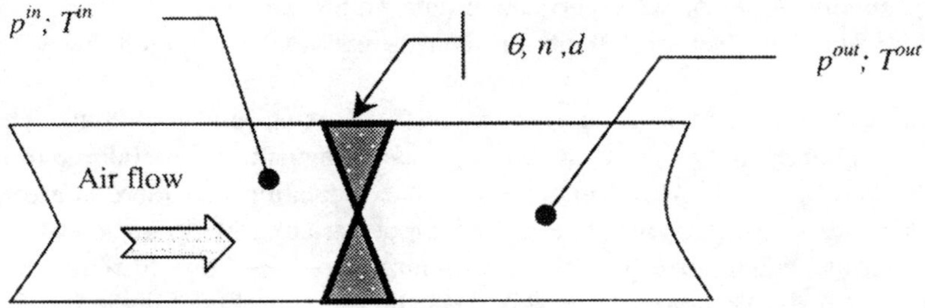

Fig. 8.8. Ducted fan

Outlet air temperature and pressure are determined by solving these two equations and related expressions for the total pressure rise across the fan and fan absorbed power (air density is a standard value, which could be supposed to be constant). Geometrical parameters of fans such as diameter, speed and blade angle (in case of axial fans) are normally ignored or represented by a polynomial *black-box* model (Brandemuehl et. al., 1998).

The normalised values for the pressure rise (ψ) and absorbed power (λ) on the normalised volume flow rate (ϕ) have been introduced as (Wright, 1991):

$$\psi = \frac{2\,p_{tf}}{\rho_i(\pi dn)^2};$$

$$\phi = \frac{4m}{\rho_i \pi^2 d^3 n};$$

$$\lambda = \frac{8q}{\rho_i \pi^4 d^5 n^3};$$

where ρ_i denotes air density at fan inlet;

d is the fan diameter;

n is the fan speed.

FRB models representing ψ and λ as functions of the blade angle (θ) and normalised volumetric flow rate (ϕ) could be generated for axial fans.

A two-speed axial fan with diameter *1m* is considered, running at *975* and *1470 rpm* respectively. A set of 62 experimental data points was taken from the literature (Wright, 1991). Part (set of 10) of the data points was considered for validation purposes.

In this example there is 3 *flexible* variables (2 inputs and 1 output). The first variable, normalised flow rate of the air (ϕ), has 6 linguistic terms (although two of them - *Rather Small* and *Medium*) are very close indicating that there is a room for model structure simplification). The second input variable, blade angle (θ) has five linguistic terms, which corresponds to the training data based on 5 different standard blade angles values. They are tabulated in the Table 8.2 and plotted in the Fig. 8.9.

The model consists of *r=6 flexible* rules:

$$R_1\text{:}\textbf{\textit{IF}}\ (\phi \text{ is Small})\ \textbf{\textit{AND}}\ (\theta \text{ is Medium})\quad \textbf{\textit{THEN}}\ \lambda = a_0^1 + a_1^1\phi + a_2^1\theta \qquad (8.4)$$

$$R_2\text{:}\textbf{\textit{IF}}\ (\phi \text{ is Very Small})\ \textbf{\textit{AND}}\ (\theta \text{ is Very Small})\quad \textbf{\textit{THEN}}\ \lambda = a_0^2 + a_1^2\phi + a_2^2\theta$$

$$R_3\text{:}\textbf{\textit{IF}}\ (\phi \text{ is Rather Small})\ \textbf{\textit{AND}}\ (\theta \text{ is Very Big})\quad \textbf{\textit{THEN}}\ \lambda = a_0^3 + a_1^3\phi + a_2^3\theta$$

$$R_4\text{:}\textbf{\textit{IF}}\ (\phi \text{ is Medium})\ \textbf{\textit{AND}}\ (\theta \text{ is Big})\quad \textbf{\textit{THEN}}\ \lambda = a_0^4 + a_1^4\phi + a_2^4\theta$$

$$R_5\text{:}\textbf{\textit{IF}}\ (\phi \text{ is Big})\ \textbf{\textit{AND}}\ (\theta \text{ is Small})\quad \textbf{\textit{THEN}}\ \lambda = a_0^5 + a_1^5\phi + a_2^5\theta$$

$$R_6\text{:}\textbf{\textit{IF}}\ (\phi \text{ is Very Big})\ \textbf{\textit{AND}}\ (\theta \text{ is Very Big})\quad \textbf{\textit{THEN}}\ \lambda = a_0^6 + a_1^6\phi + a_2^6\theta$$

Table 8.2. Linguistic labels of *flexible* variables (θ ,ϕ) for the model of the Fan Power

Blade Angle	**Flow rate**
Very Small (VS)	*Very Low (VL)*
Small (S)	*Low (L)*
Medium (M)	*Rather Low (RL)*
Big (B)	*Medium (M)*
Very Big (VB)	*Big (B)*
	Very Big (VB)

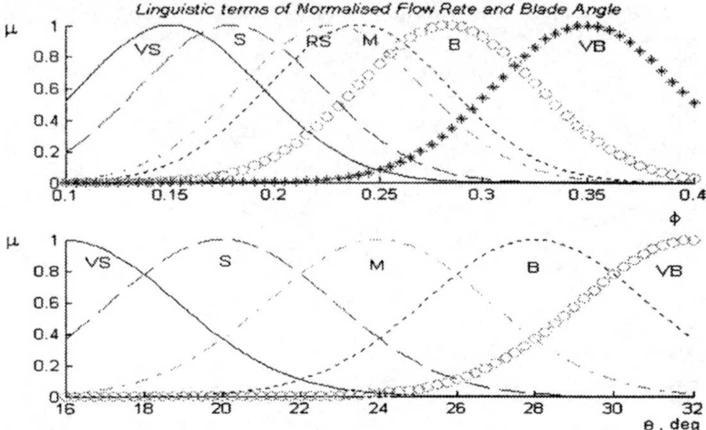

Fig. 8.9. Linguistic terms of the *flexible* variables Normalised Flow Rate and Blade Angle for the model of the Fan Power

Product moment correlation coefficient in validation of the best FRB model of the fan total power is 0.9991 and the root mean squares error is 0.00037 (Fig. 8.10).

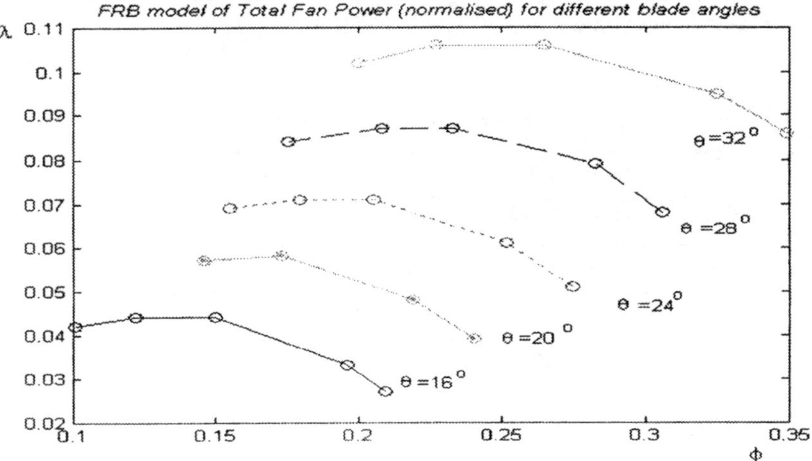

Fig. 8.10. Normalised fan absorbed power as a function of normalised volumetric flow

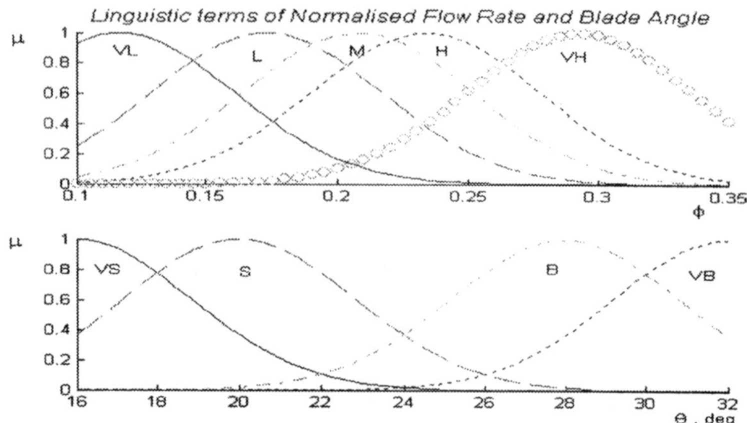

Fig. 8.11. Linguistic terms of the *flexible* variables Normalised Flow Rate and Blade Angle for the model of the Fan Pressure

The model of the absorbed pressure has also 3 *flexible* variables (the same two inputs and the output - normalised value of the absorbed fan pressure). The normalised flow rate of the air (ϕ), however, has 5 linguistic terms for this model. The second input variable, blade angle (θ) has four linguistic terms tabulated in the Table 8.3. The reason smaller number of linguistic terms is necessary for this model (Fig. 8.11) is, probably, that the curves are closer to each other (Fig. 8.12).

Table 8.3. Linguistic labels of *flexible* variables (θ, ϕ) for the model of the Fan Pressure

Blade angle	**Flow rate**
Very Small	*Very Low*
Small	*Low*
Big	*Medium*
Very Big	*High*
	Very High

The FRB model has correlation 0.9992 and root mean squares 0.0042 (Fig. 8.12) and consists of *r=6 flexible* rules:

R_1:**IF** (ϕ *is Medium*) **AND** (θ *is Big*) **THEN** $\psi = b_0^1 + b_1^1\phi + b_2^1\theta$ (8.5)

R_2:**IF** (ϕ *is Very High*) **AND** (θ *is Big*) **THEN** $\psi = b_0^2 + b_1^2\phi + b_2^2\theta$

R_3:**IF** (ϕ *is High*) **AND** (θ *is Small*) **THEN** $\psi = b_0^3 + b_1^3\phi + b_2^3\theta$

R_4:**IF** (ϕ *is Very Low*) **AND** (θ *is Very Small*) **THEN** $\psi = b_0^4 + b_1^4\phi + b_2^4\theta$

R_5:**IF** (ϕ *is Very High*) **AND** (θ *is Very Big*) **THEN** $\psi = b_0^5 + b_1^5\phi + b_2^5\theta$

R_6:**IF** (ϕ *is Low*) **AND** (θ *is Small*) **THEN** $\psi = b_0^6 + b_1^6\phi + b_2^6\theta$

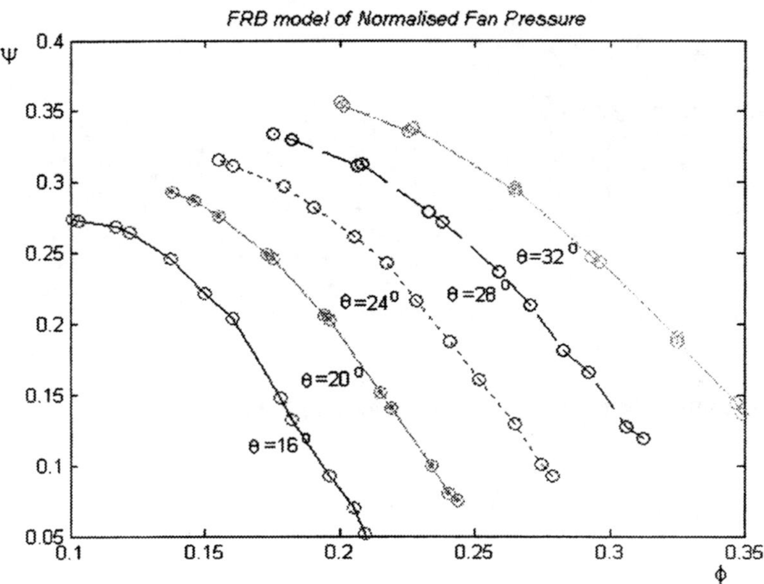

Fig. 8.12. Normalised fan pressure rise (ψ) as a function of normalised volumetric flow rate (ϕ).

8.1.3
Modelling Efficiency of Boilers: Hybrid Model Approach

In this section the problem of modelling small nominally rated 300kW, natural gas fired boiler is considered. It could be used to supply medium temperature hot water to the HVAC. Training data covered a typical range of operation, which is assumed to operate at a constant water mass flow rate and a flow and return water temperature difference of 15 degree K (Angelov et. al., 2000c).

Control of boiler operation is typically based on the return water temperature. Firings rate and return water temperature was considered to be the model inputs. They have been considered to vary between the maximum firing rate down to 10% of that rate, and for return water temperatures between 20°C and 100°C.

A hybrid scheme, which consists of conventional (first principle-based), and FRB models have been considered (Fig. 8.13). The resulting model is suitable for normal operating conditions, and for "start-up" operation with the heating fluid at ambient temperatures. The water mass flow rate through the boiler is supposed to be constant at 3.8kg/s.

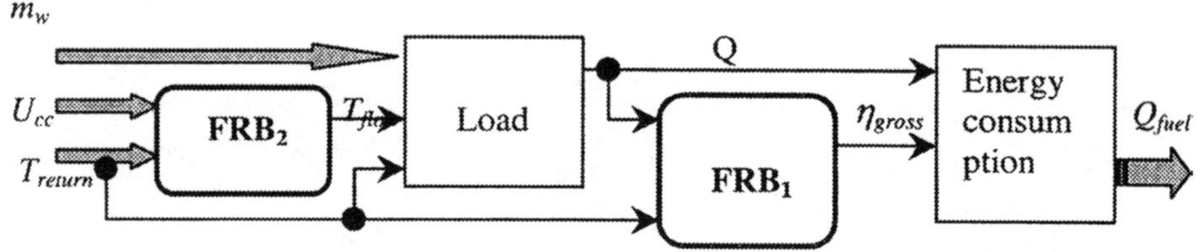

Fig. 8.13. Hybrid Model of a sub-system including a boiler

In this scheme, the conventional (first principle) models are given in rectangles:

✓ Load model :

$$Q = \dot{m}_w Cp_w (T_{flow} - T_{retrum}), \tag{8.6}$$

✓ Fuel consumption model:

$$Q_{fuel} = \frac{Q}{\eta_{gross}} \tag{8.7}$$

where Q is the load across the boiler,

$\dot{m}_w$ and Cp_w are the mass flow rate and specific heat capacity of water;

T_{flow} and T_{retrum} are the flow and return water temperatures respectively;

Q_{fuel} is the instantaneous fuel consumption, *kJ/s*;

η_{gross} is the gross efficiency of the boiler.

Two *flexible* rule-based models (FRB_1 and FRB_2) are considered. They were generated from the data. The first predicts the gross efficiency of the boiler (η_{gross}), as a function of the boiler load (Q) and return water temperature (T_{return}). The second modelled the flow temperature (T_{flow}) taking the return water temperature (T_{return}) and the control signal (U_{cc}) to the boiler as inputs. Control signal is a percent of the maximum firing rate.

The second model allows for incorporation of the component into a HVAC sub-system performance simulation. The first one can be used in conjunction with this to generate predictions of fuel consumption (Q_{fuel}) for energy analysis, like annual energy cost predictions.

In each case 7 linguistic terms and less than $R=20$ rules where used to characterise the process. The product-moment-correlation-coefficients between the measured and the experimental values where:

- Training
  - Gross efficiency: 0.982;
  - Flow temperature: 0.990;
- Validation:
  - Gross efficiency: 0.986;
  - Flow temperature: 0.988.

Both models have quite small errors and very high correlation. The efficiency model was generated at ~2000 generations and the flow temperature model at ~1000. Both FRB models and the overall hybrid model produced quite smooth input/output surfaces.

8.2 Modelling the Thermal Load of a Building

A problem of modelling thermal load of a building is considered as another application. It is an important component of the optimal design of a building aiming minimisation of energy consumed while keeping comfort in it.

The goal is to model heat load (Q) as a function of outdoor temperature (T^{amb}), wind speed (W), solar radiation (I), time of the day (t) and occupants activity level (A). It could be represented as a FRB model. Linguistic terms representing the *flexible* variables are tabulated in the Table 8.4.

Total number of possible *flexible* rules in this problem is $TNR=5^6=15625$. Instead of using such big chromosome $R=20$ genes are considered, which determines the upper boundary of the number of rules expected to be in the final model.

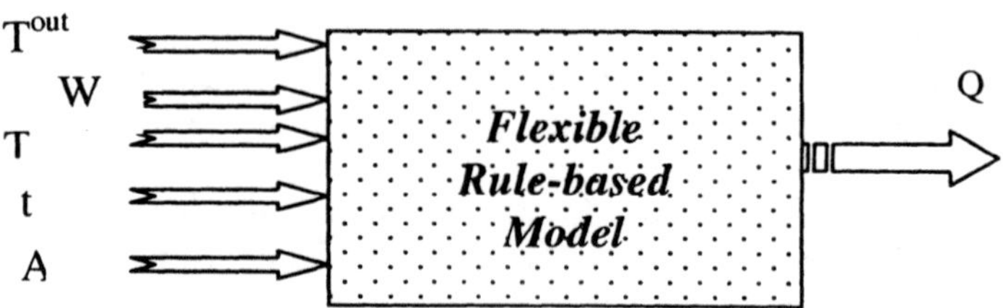

Fig. 8.14. FRB model of the thermal load of a building

Table 8.4. Linguistic labels of *flexible* variables (I, T^{amb}, A, t, W)

Solar radiation, Temperature, Occupants' activity	Time of the day	Wind speed
Very Low (VL)	*Early Morning (EM)*	*Very Strong (VS)*
Low (L)	*Morning (Mo)*	*Strong (S)*
Medium (M)	*Noon (Nn)*	*Medium (M)*
High (H)	*Evening (E)*	*Mild (Mld)*
Very High (VH)	*Night (N)*	*Very Mild (VM)*

As a result a model with smaller ($r<R$) number of rules is found due to the coincidence of some indices and zeros appearing as index values. Another 30 genes represent centres of the Gaussian membership functions. Spreads are fixed to be equal:

$$\sigma_{ij}=0.1\frac{\overline{x_l}-\underline{x_l}}{m_l}c_{lj}\,;\; l=1,2,...,5;\; j=1,2,...,5.$$

The output is represented by triangular membership functions.

I_1	I_2	...	I_{20}	c_{11}	c_{12}	...	c_{15}	...	c_{61}	c_{62}	...	c_{65}

Fig. 8.15. Genotype - Thermal load modelling problem

A set of 20 samples is considered (Angelov, 1999). A part of them is used for learning while the rest (Table 8.6) are left for the validation test. The lower and upper bounds of the linguistic variables are tabulated in the Table 8.5.

Table 8.5. Boundaries of linguistic variables

	T, °C	W, *m/s*	I, *W/m*^{2}	t, *h*	A, -	Q, *kW*
$\underline{x}$	*0*	*0*	*0*	*0*	*Very Low*	*0*
$\overline{x}$	*30*	*40*	*10*	*23*	*Very High*	*10*

Table 8.6. Data used for validation

experiment	T, °C	W, *m/s*	I, *W/m*^{2}	t, *h*	A, -	Q, *kW*
1	*12*	*20*	2	*8*	*Very Low*	*8*
2	26	*5*	*9*	*14*	*High*	*3*
3	*20*	*10*	*5*	*11*	*Low*	*4*
4	*17*	*15*	*3*	*10*	*Low*	*6*
5	*19*	*10*	*8*	*16*	*Medium*	*3*

The problem of structure and parameter identification is formulated as follows:

$$J=\sum_{i=1}^{N}(Q-\hat{Q})^2 \rightarrow \min \qquad (8.8)$$

subject to: *FRB model*

$1 \leq I_i \leq 15625; \qquad i=1,2,...,20; \quad j=1,2,...,5$

$$(j-1)\frac{\overline{x}-\underline{x}}{6} \leq c_{lj} \leq (j+1)\frac{\overline{x}-\underline{x}}{6}$$

$\sigma_{3j}=\sigma_{5j}=1;$

$\sigma_{1j}=10;$

$\sigma_{2j}=4;$

$\sigma_{4j}=2.4;$

An initial population of a number of individuals (in this case population includes 30 chromosomes) each of them consisting of 50 genes is generated randomly to satisfy constraints in (8.8). After an evolution driven by selection of the fittest a set of *flexible* rules is generated which describes validation data with an average error of less *than 0.2 kW* (Angelov, 1999).

The following chromosome gives the best individual:

Fig. 8.16. The genotype of the best chromosome (using rules' indices encoding)

Rules' indices									
R_1	R_2	R_3	R_4	R_5	R_6	R_7	R_8	R_9	R_{10}
3208	10436	13850	5498	5249	3979	15141	4857	10436	14608

Wind Speed; *W, m/s*					Ambient Temp.;T^{amb}, ^oC					Solar Radiation; *I, W/m*^{2}				
VL	*L*	*M*	*H*	*VH*	*VL*	*L*	*M*	*H*	*VH*	*VL*	*L*	*M*	*H*	*VH*
5.4	6.7	12	16.1	16.8	11.1	14.5	18.4	27.9	28	1.6	3.9	4.1	8	8.2

Time of the day; *t, h*					Occupant(s)' Activity; *A, %*					Heat Load; *Q, kW*				
EM	*Mo*	*Nn*	*E*	*N*	*VL*	*L*	*M*	*H*	*VH*	*VL*	*L*	*M*	*H*	*VH*
3.8	4.8	11.3	16.3	22.9	2.7	22.4	58.1	69.4	83.7	1.7	3.3	5	6.7	8.3

Applying decoding procedure (Section 4, Chapter 4) this chromosome is transformed into the following set of *flexible* rules:

R_1: ***IF (T^{amb} is High) AND (W is Strong) AND (I is Low) AND (t is Morning) AND (A is Low) THEN (Q is High)***

R_2: ***IF (T^{amb} is High) AND (W is Mild) AND (I is High) AND (t is Noon) AND (A is Medium) THEN (Q is Low)***

R_3: ***IF (T^{amb} is Medium) AND (W is Medium) AND (I is Medium) AND (t is Noon) AND (A is Low) THEN (Q is Medium)***

R_4: ***IF (T^{amb} is High) AND (W is Very Strong) AND (I is Low) AND (t is Morning) AND (A is High) THEN (Q is Low)***

R_5: ***IF (T^{amb} is Low) AND (W is Mild) AND (I is Low) AND (t is Night) AND (A is Very Low) THEN (Q is Very High)***

R_6: ***IF (T^{amb} is Low) AND (W is Strong) AND (I is Low) AND (t is Night) AND (A is Very High) THEN (Q is Very High)***

R_7: ***IF (T^{amb} is Low) AND (W is Medium) AND (I is Medium) AND (t is Night) AND (A is Very High) THEN (Q is High)***

R_8: ***IF (T^{amb} is High) AND (W is MIld) AND (I is High) AND (t is Morning) AND (A is Very Low) THEN (Q is Very Low)***

R_9: ***IF (T^{amb} is High) AND (W is Medium) AND (I is High) AND (t is Morning) AND (A is Low) THEN (Q is Low)***

R_{10}: ***IF (T^{amb} is Medium) AND (W is Medium) AND (I is Low) AND (t is Morning) AND (A is Very Low) THEN (Q is High)***

Applying the second encoding approach, which is based on the linguistic terms (Section 4, Chapter 4) the same result is achieved (Fig. 8.16a illustrates only the first part of the chromosome, the second part, consisting of the parameters of linguistic terms is the same as in the Fig. 8.16):

R_1						R_2						R_3						R_4						R_5					
T	*W*	*I*	*t*	*A*	*Q*	*T*	*W*	*I*	*t*	*A*	*Q*	*T*	*W*	*I*	*t*	*A*	*Q*	*T*	*W*	*I*	*t*	*A*	*Q*	*T*	*W*	*I*	*t*	*A*	*Q*
4	3	1	0	1	3	3	1	3	2	2	1	2	2	2	2	1	2	4	4	0	1	3	1	1	1	1	4	0	4

R_1						R_2						R_3						R_4						R_5					
T	*W*	*I*	*t*	*A*	*Q*	*T*	*W*	*I*	*t*	*A*	*Q*	*T*	*W*	*I*	*t*	*A*	*Q*	*T*	*W*	*I*	*t*	*A*	*Q*	*T*	*W*	*I*	*t*	*A*	*Q*
1	3	1	4	4	4	1	3	3	4	4	3	4	1	4	1	0	0	3	2	3	1	1	1	3	3	0	1	0	4

Fig. 8.16a. The genotype of the best chromosome (using linguistic terms encoding)

Linguistic terms are represented in the Fig. 8.17 - Fig. 8.19. It is interesting to mention that the linguistic variables *Solar radiation* (*I*) and *Wind Speed* (*W*) could be represented by three linguistic terms only (Figs. 8.17 and 8.18) due to the coincidence or closeness of the values of the parameters of linguistic terms. .

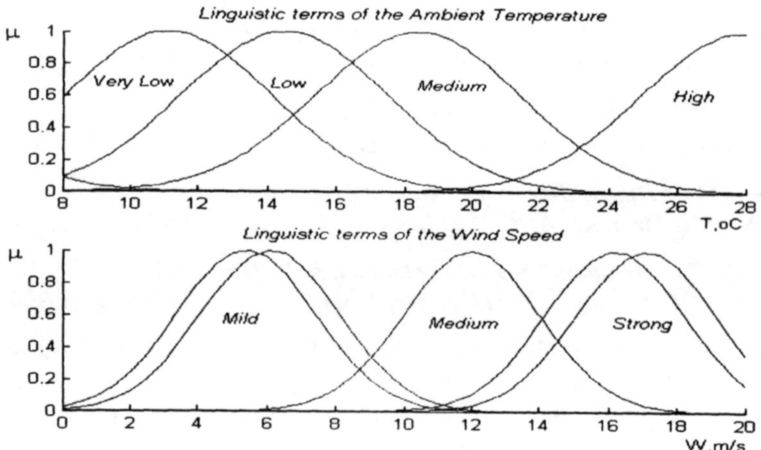

Fig. 8.17. *Fuzzy* linguistic terms of Ambient temperature and Wind speed

Similarly, the linguistic variables *Time of the Day* (t) and *Ambient Temperature* (T) could be represented by four linguistic variables only. For the *Time of the Day* they are:

- *Morning;*
- *Noon;*
- *Evening;*
- *Night*

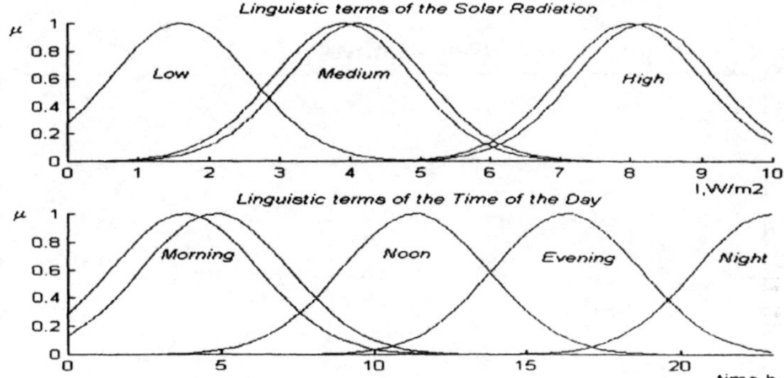

Fig. 8.18. *Fuzzy* linguistic terms of Solar radiation and Time of the day

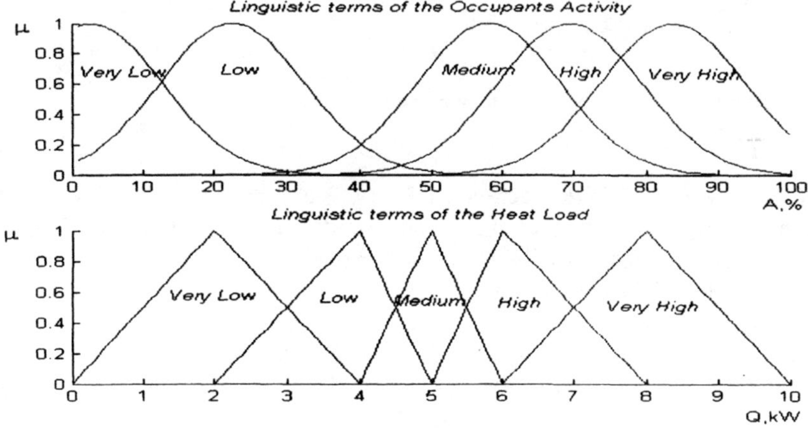

Fig. 8.19. *Fuzzy* linguistic terms of Occupants' activity and Heat load

8.3
Learning trough Experience (VL Strategy)

The same *real* data (courtesy of ASHRAE for the use of data, generated from the ASHRAE funded research project RP1020) from the section 8.1.1.1 have been used. The measurements of T_a^{in}, m_a, and U_{cc} formed inputs to the model. The output is the outlet of the coil air temperature. Samples were taken at one-minute intervals for two 24-hour periods, a few days apart.

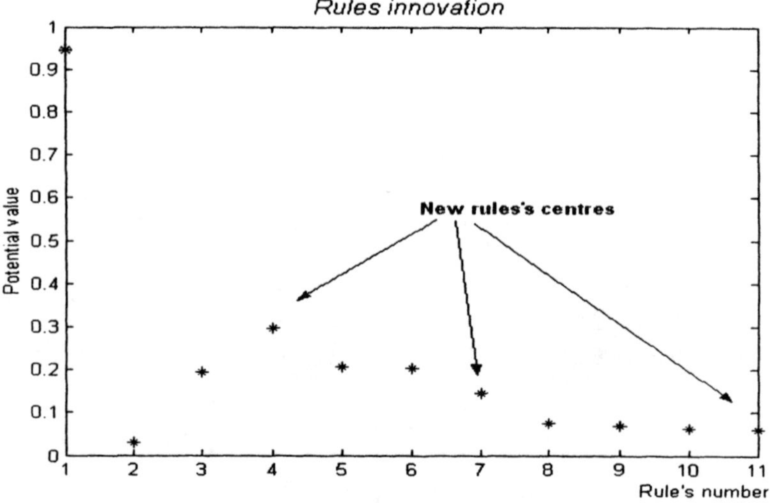

Fig. 8.20. Rules *innovation* represented by potential values

The RH strategy applied in the section 8.1.1.1 (where a relatively high (N=*500*) set of data have been considered) is compared to the VL strategy when only N=*10* data samples have been used initially in *off-line* mode. The model structure has been practically developed throughout the process in *on-line* mode.

Only one rule was generated based on *off-line* data (upper left corner of the Fig.. 8.20). The ***e*R** model, however, generates ten new rules based on potentials of new data points (Fig. 8.21) and succeeds in keeping the error low. The potentials before subtracting penalty for closeness are depicted in the Fig. 8.22.

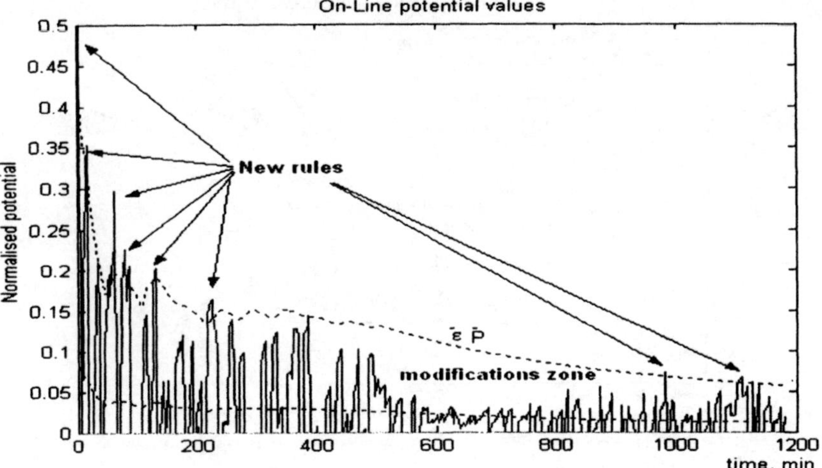

Fig. 8.21. *On-line* up-dated potentials

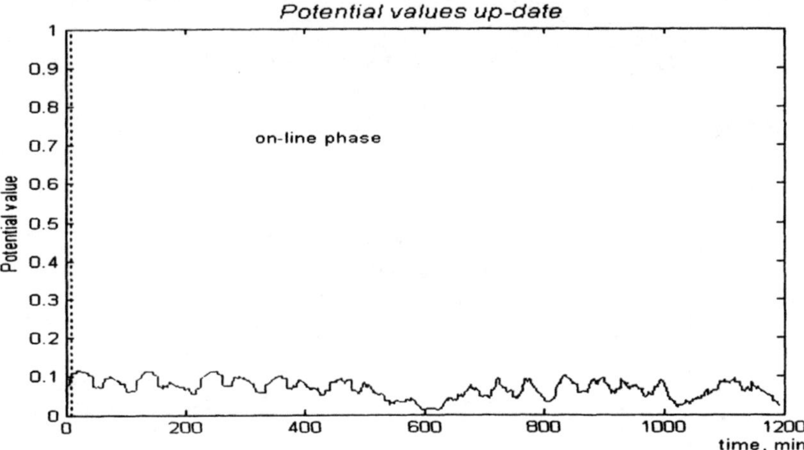

Fig. 8.22. Potential values of all data samples before subtraction

The new rules generated in *on-line* mode has different linguistic terms, parameters (centres) of which for the inlet air temperature and the control signal to the valve are presented in the Fig. 8.23 and Fig. 8.24 respectively. Some of them could be redundant and the model could be simplified using GA, as will be discussed later.

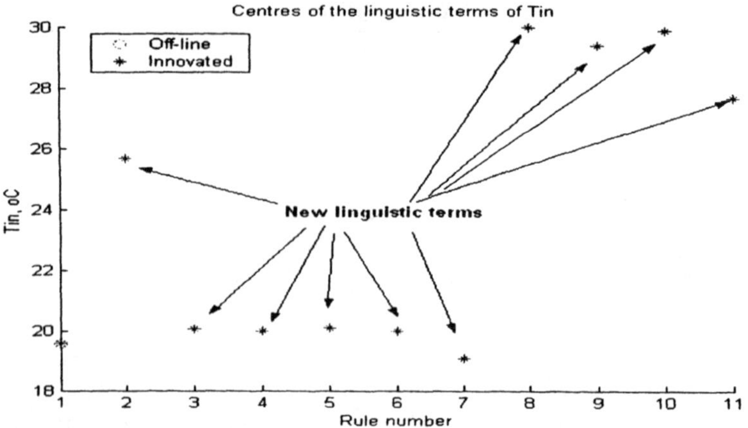

Fig. 8.23. Linguistic terms' parameters (Tin)

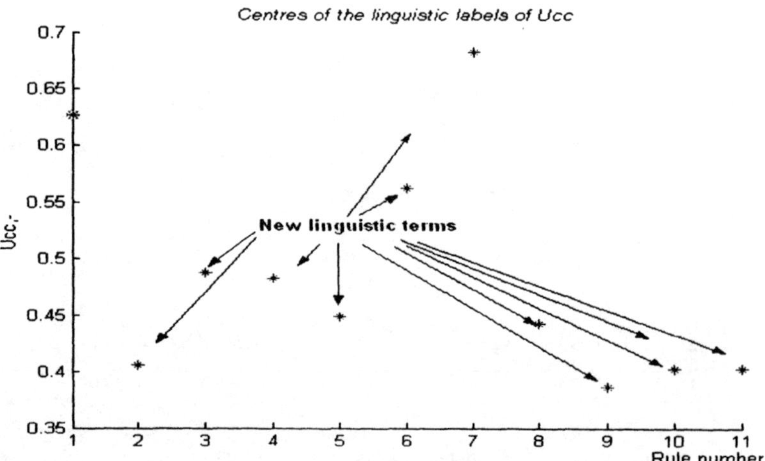

Fig. 8.24. Linguistic terms' parameters (U_{cc})

As a result of this permanent up-date of both the model structure and parameters performance of the ***e*R** model is significantly better than that of the *off-line* trained FRB model. As seen in the Fig. 8.25 for the absolute error (the root mean error for the

***e*R** model is 0.0021 degree K versus 0.018 degree K for the *off-line* trained FRB model) and in the Fig. 8.26 for the original and model prediction of the outlet temperature, the *off-line* trained FRB model progressively worsen its performance (most significantly after the *10*[th] hour (*600*[th] min)), while the error of the ***e*R** model is reasonably low.

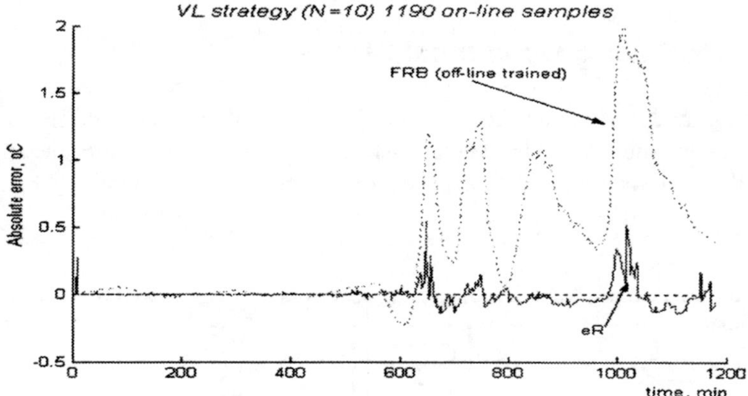

Fig. 8.25. Prediction error of control signal by eR and *off-line* FRB model

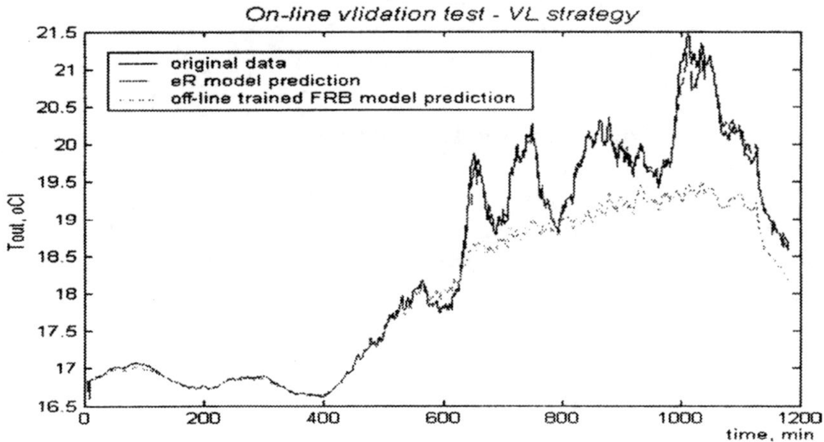

Fig. 8.26. *On-line* validation test (original data vs. predictions by eR and FRB models)

This is an illustration of the real *learning trough experience* of ***e*R** models evolving both their structure and parameters in the process of collecting the data. It should be noted that the time necessary for calculation is less than a second using 550MHz PC, 128 MB RAM for each new data sample as the procedure is ***non-iterative*** and ***recursive***. This test (using limited initial information) illustrates the potential of the approach to be implemented in robotics.

8.4 *On-line* Modelling Dynamical Signals

A regression model of the control signal (U_{cc}) to the valve controlling the water flow rate to a fan-coil sub-system has been considered as a demonstration of the application of ***e*R** models to dynamical problems. The test system is shown in the Fig. 8.27.

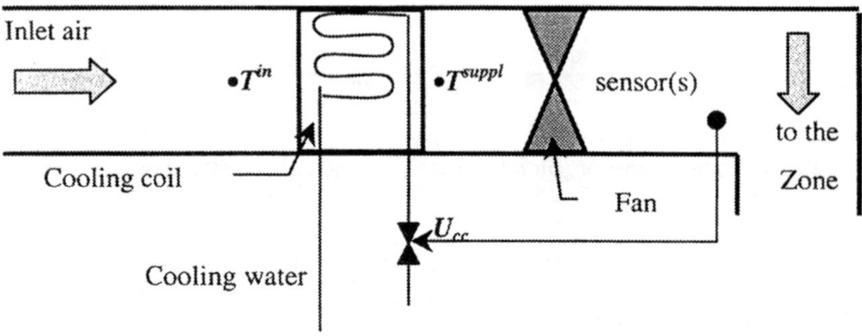

Fig. 8.27. Test fan-coil sub-system of an HVAC system

Training data used in the problem (U_{cc}, T^{in}, T^{suppl}) are shown in the Figs. 8.28-8.30. The data was collected from a *real* system (courtesy of ASHRAE for the use of data, generated from the ASHRAE funded research project RP1020).

The plot in the Fig. 8.28 shows the control signal at the present moment and $\Delta K = 20$ steps ahead. The present value of U_{cc} ($U_{cc}|_k$) is one of the inputs of the FRB model considered, while the predicted value 20 steps ahead is its output (8.9). The other inputs are the present and past (one step back) values of the air inlet (T^{in}) and supply to the zone (T^{suppl}) temperature:

$$U_{cc}\big|_{k+\Delta K} = FRB\left(U_{cc}\big|_{k}; T^{in}\big|_{k}; T^{\sup\, pl}\big|_{k}; U_{cc}\big|_{k-1}; T^{in}\big|_{k-1}; T^{\sup\, pl}\big|_{k-1}\right) \tag{8.9}$$

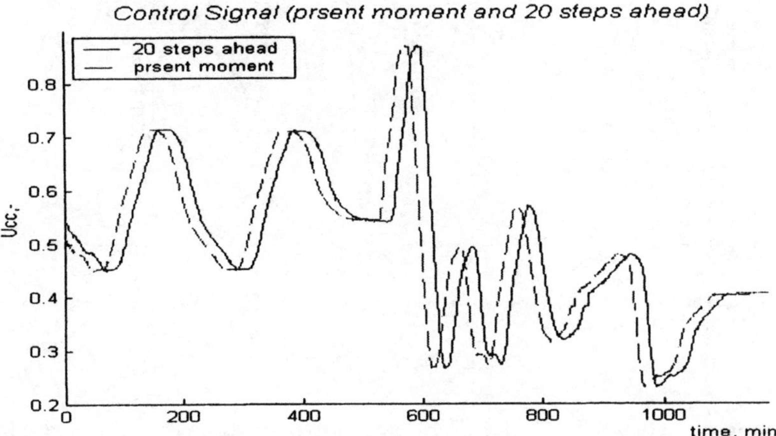

Fig. 8.28. Control signal to the valve (present moment and 20 steps ahead prediction)

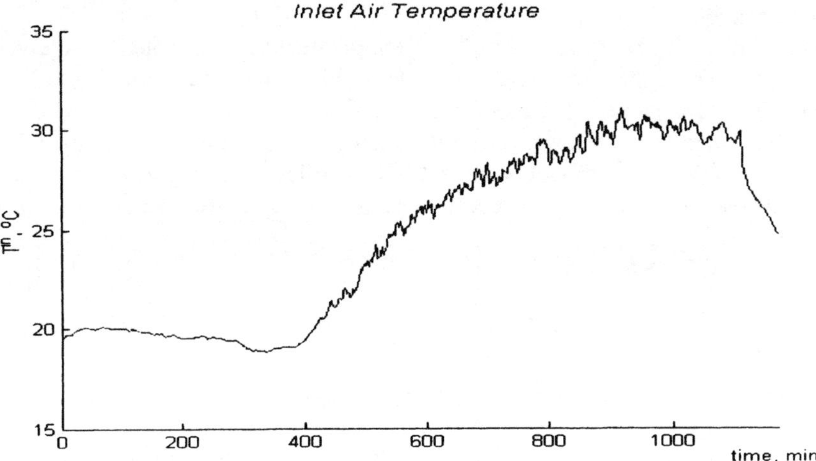

Fig. 8.29. Air inlet temperature

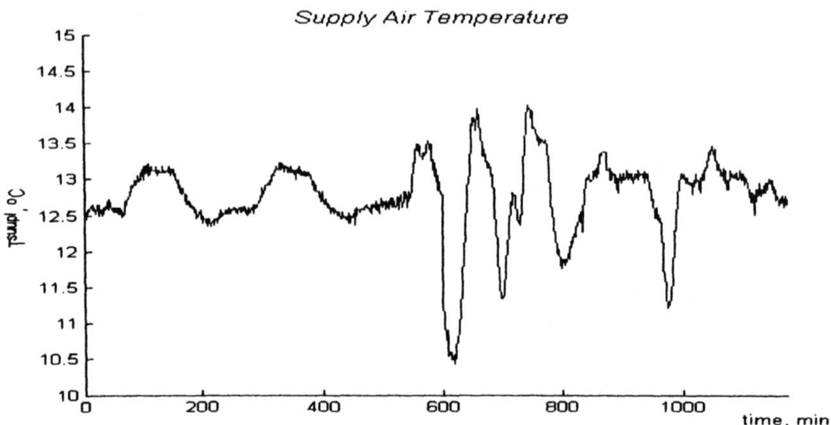

Fig. 8.30 Supply air temperature

The present values of T^{in} and T^{suppl} are shown in the Fig. 8.29 and Fig. 8.30 respectively.

The coil cools the warm air that flows on. The cool air is used to maintain comfortable conditions in an occupied Zone. One of the principle loads on the coil is generated due to the supply of ambient air required to maintain a minimum standard of indoor air quality.

The inlet (T^{in}) and supply air (T^{suppl}) temperatures are sampled at intervals of 1 minute. Data from the same *real* sub-system, but from a different day (19th August 1998) was used to validate the models.

On-line identification approach to the FRB model, as discussed in the Chapter 7, has been applied. ***e*R** model with 7 *flexible* rules and 7 membership functions describing each input variable has been generated based on the data in *real time* mode:

$$R_1\text{: } \textbf{\textit{IF}}\ (U_{cc}|_k \text{ is } H)\ \textbf{\textit{AND}}\ (U_{cc}|_{k-1} \text{ is } H)\ \textbf{\textit{AND}}\ (T^{in}|_k \text{ is } VL)\ \textbf{\textit{AND}}\ (T^{in}|_{k-1} \text{ is } VL)\ \textbf{\textit{AND}}\ (T^{\sup pl}|_k \text{ is } L)\ \textbf{\textit{AND}}\ (T^{\sup pl}|_{k-1} \text{ is } L)\quad \textbf{\textit{THEN}}\ U_{cc}|_{k+\Delta K} = f_1$$

$$R_2\text{: } \textbf{\textit{IF}}\ (U_{cc}|_k \text{ is } VL)\ \textbf{\textit{AND}}\ (U_{cc}|_{k-1} \text{ is } VL)\ \textbf{\textit{AND}}\ (T^{in}|_k \text{ is } VH)\ \textbf{\textit{AND}}\ (T^{in}|_{k-1} \text{ is } VH)\ \textbf{\textit{AND}}\ (T^{\sup pl}|_k \text{ is } EH)\ \textbf{\textit{AND}}\ (T^{\sup pl}|_{k-1} \text{ is } EH)\quad \textbf{\textit{THEN}}\ U_{cc}|_{k+\Delta K} = f_2$$

R_3: **IF** $(U_{cc}|_k$ is *EH*) **AND** $(U_{cc}|_{k-1}$ is *EH*) **AND** $(T^{in}|_k$ *is EL*) **AND** $(T^{in}|_{k-1}$ *is EL*) **AND** $(T^{\sup pl}|_k$ *is H*) **AND** $(T^{\sup pl}|_{k-1}$ *is H*) **THEN** $U_{cc}|_{k+\Delta K}=f_3$

R_4: **IF** $(U_{cc}|_k$ is *VH*) **AND** $(U_{cc}|_{k-1}$ is *VH*) **AND** $(T^{in}|_k$ *is L*) **AND** $(T^{in}|_{k-1}$ *is L*) **AND** $(T^{\sup pl}|_k$ *is VL*) **AND** $(T^{\sup pl}|_{k-1}$ *is VL*) **THEN** $U_{cc}|_{k+\Delta K}=f_4$

R_5: **IF** $(U_{cc}|_k$ is *EL*) **AND** $(U_{cc}|_{k-1}$ is *EL*) **AND** $(T^{in}|_k$ *is EH*) **AND** $(T^{in}|_{k-1}$ *is EH*) **AND** $(T^{\sup pl}|_k$ *is H*) **AND** $(T^{\sup pl}|_{k-1}$ *is H*) **THEN** $U_{cc}|_{k+\Delta K}=f_5$

R_6: **IF** $(U_{cc}|_k$ is *L*) **AND** $(U_{cc}|_{k-1}$ is *L*) **AND** $(T^{in}|_k$ *is M*) **AND** $(T^{in}|_{k-1}$ *is M*) **AND** $(T^{\sup pl}|_k$ *is H*) **AND** $(T^{\sup pl}|_{k-1}$ *is H*) **THEN** $U_{cc}|_{k+\Delta K}=f_6$

R_7: **IF** $(U_{cc}|_k$ is *M*) **AND** $(U_{cc}|_{k-1}$ is *M*) **AND** $(T^{in}|_k$ *is H*) **AND** $(T^{in}|_{k-1}$ *is H*) **AND** $(T^{\sup pl}|_k$ *is EL*) **AND** $(T^{\sup pl}|_{k-1}$ *is WL*) **THEN** $U_{cc}|_{k+\Delta K}=f_7$

The following notations has been used for convenience:

EL for *Extremely Low*
VL for *Very Low*
L for *Low*
M for *Medium*
H for *High*
VH for *Very High*
EH for *Extremely High*

Linguistic terms of the supply air temperature for the present and past moments are presented in the Fig. 8.31a.

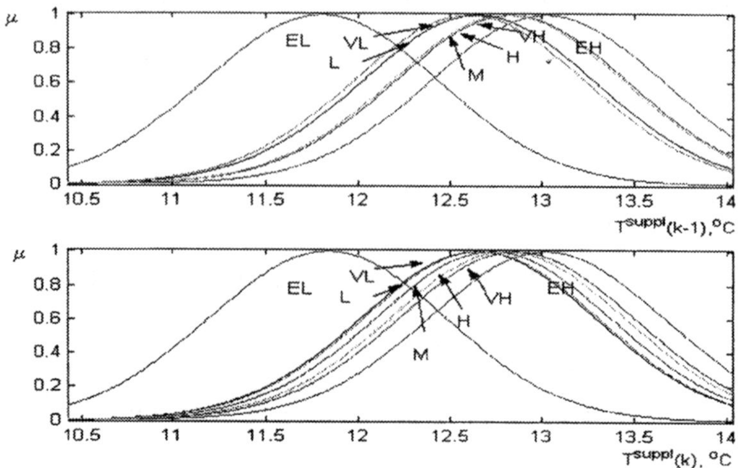

Fig. 8.31a. Linguistic terms of the supply air temperature

Linguistic terms (and respective membership functions) describing the inlet air temperature are given in the Fig. 8.32a.

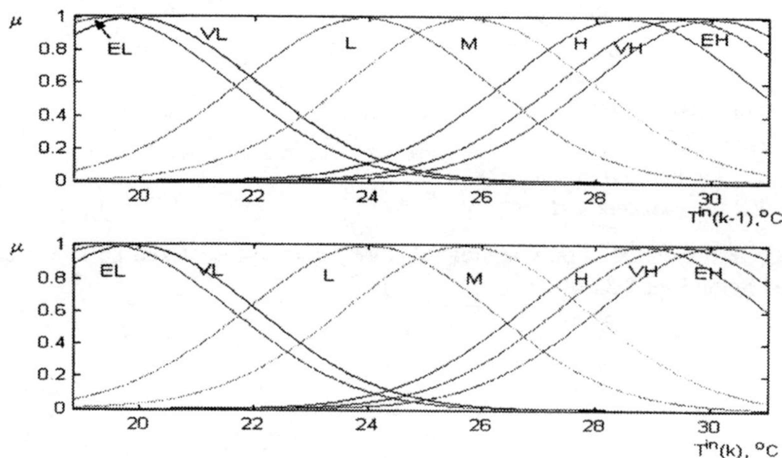

Fig. 8.32a. Linguistic terms of the inlet air temperature

Fig. 8.33a represents the linguistic terms for the control signal at the past and the present moment.

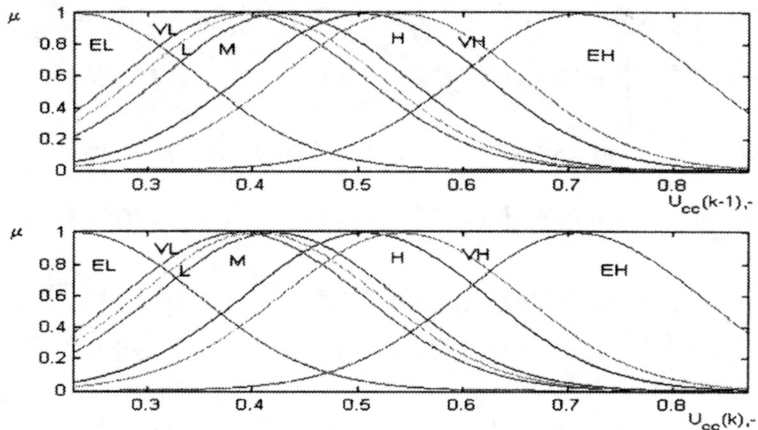

Fig. 8.33a. Linguistic terms of the control signal

8.5 Model Simplification by Linguistic Term's Reduction

The number of membership functions generated *on-line* by the quasi-linear approach for this test case is *42* (*7* linguistic terms for *6 flexible* variables). Applying the reduction of linguistic terms as discussed in Sections 5.4.1 and 7.7.1, a significant simplification of this (initial) model could be achieved non-iteratively in *on-line* mode.

The simplified model is:

$$R_1\text{: } \textbf{\textit{IF}}\ (U_{cc}|_k \text{ is } M)\ \textbf{\textit{AND}}\ (U_{cc}|_{k-1} \text{ is } M)\ \textbf{\textit{AND}}\ (T^{in}|_k \text{ is } VL)\ \textbf{\textit{AND}}\ (T^{in}|_{k-1} \text{ is } VL)\ \textbf{\textit{AND}}\ (T^{\sup pl}|_k \text{ is } L)\ \textbf{\textit{AND}}\ (T^{\sup pl}|_{k-1} \text{ is } L)\quad \textbf{\textit{THEN}}\ U_{cc}|_{k+\Delta K} = f_1$$

$$R_2\text{: } \textbf{\textit{IF}}\ (U_{cc}|_k \text{ is } L)\ \textbf{\textit{AND}}\ (U_{cc}|_{k-1} \text{ is } L)\ \textbf{\textit{AND}}\ (T^{in}|_k \text{ is } H)\ \textbf{\textit{AND}}\ (T^{in}|_{k-1} \text{ is } H)\ \textbf{\textit{AND}}\ (T^{\sup pl}|_k \text{ is } L)\ \textbf{\textit{AND}}\ (T^{\sup pl}|_{k-1} \text{ is } L)\quad \textbf{\textit{THEN}}\ U_{cc}|_{k+\Delta K} = f_2$$

R_3: **IF** $(U_{cc}\big|_k$ is $H)$ **AND** $(U_{cc}\big|_{k-1}$ is $H)$ **AND** $(T^{in}\big|_k$ *is* $VL)$ **AND** $(T^{in}\big|_{k-1}$ *is* $VL)$
AND $(T^{\sup pl}\big|_k$ *is* $L)$ **AND** $(T^{\sup pl}\big|_{k-1}$ *is* $L)$ **THEN** $U_{cc}\big|_{k+\Delta K} = f_3$

R_4: **IF** $(U_{cc}\big|_k$ is $M)$ **AND** $(U_{cc}\big|_{k-1}$ is $M)$ **AND** $(T^{in}\big|_k$ *is* $L)$ **AND** $(T^{in}\big|_{k-1}$ *is* $L)$
AND $(T^{\sup pl}\big|_k$ *is* $L)$ **AND** $(T^{\sup pl}\big|_{k-1}$ *is* $L)$ **THEN** $U_{cc}\big|_{k+\Delta K} = f_4$

R_5: **IF** $(U_{cc}\big|_k$ is $VL)$ **AND** $(U_{cc}\big|_{k-1}$ is $VL)$ **AND** $(T^{in}\big|_k$ *is* $H)$ **AND** $(T^{in}\big|_{k-1}$ *is* $H)$
AND $(T^{\sup pl}\big|_k$ *is* $L)$ **AND** $(T^{\sup pl}\big|_{k-1}$ *is* $L)$ **THEN** $U_{cc}\big|_{k+\Delta K} = f_5$

R_6: **IF** $(U_{cc}\big|_k$ is $L)$ **AND** $(U_{cc}\big|_{k-1}$ is $L)$ **AND** $(T^{in}\big|_k$ *is* $M)$ **AND** $(T^{in}\big|_{k-1}$ *is* $M)$
AND $(T^{\sup pl}\big|_k$ *is* $L)$ **AND** $(T^{\sup pl}\big|_{k-1}$ *is* $L)$ **THEN** $U_{cc}\big|_{k+\Delta K} = f_6$

R_7: **IF** $(U_{cc}\big|_k$ is $L)$ **AND** $(U_{cc}\big|_{k-1}$ is $L)$ **AND** $(T^{in}\big|_k$ *is* $H)$ **AND** $(T^{in}\big|_{k-1}$ *is* $H)$
AND $(T^{\sup pl}\big|_k$ *is* $L)$ **AND** $(T^{\sup pl}\big|_{k-1}$ *is* $L)$ **THEN** $U_{cc}\big|_{k+\Delta K} = f_7$

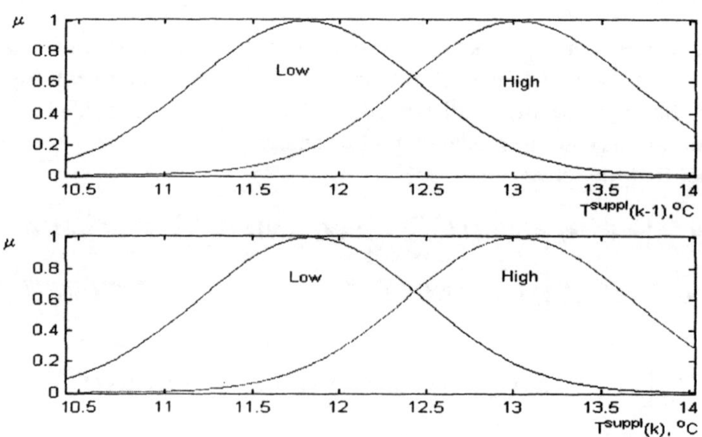

Fig. 8.31b. Linguistic terms (simplified version) of the supply air temperature

It is interesting to mention that there is a symmetry in respect to the input values of the same nature but taken at different (shifted one step) time instance. It is normal as the structure of the data has the same pattern for the same variable.

The membership functions of the simplified linguistic terms are presented in the Figs. 8.31b-8.33b

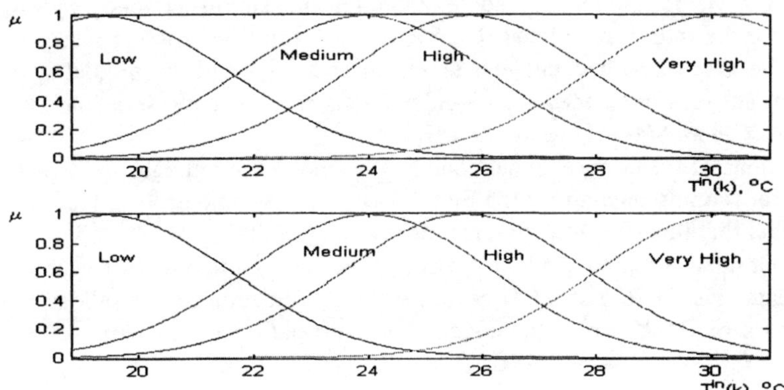

Fig. 8.32b. Linguistic terms (simplified version) of the inlet air temperature

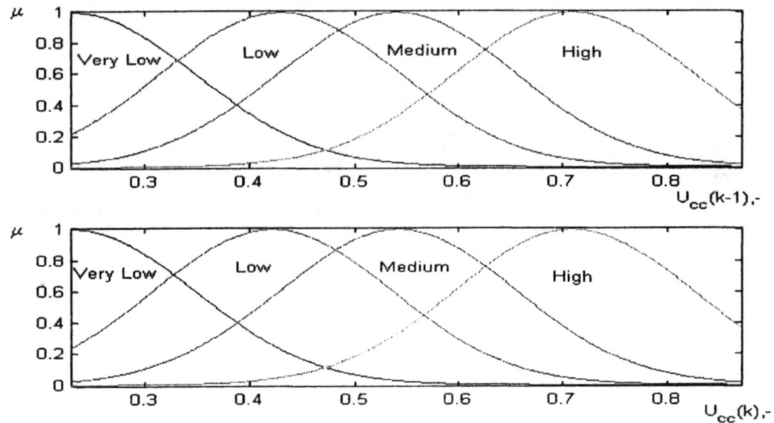

Fig. 8.33b. Linguistic terms (simplified version) of the control signal

The reduction is based on the similarity and eliminates the linguistic terms, which are close to each other (although the rules formed by them could not be so close; this has been shown in general terms in the Fig. 5.6).

As a result the number of membership functions needed is reduced to 20, as shown in the Figs. 8.31b-8.33b.It is interesting to mention that the precision does not suffer by this model simplification.

The correlation between the model predictions and the data for the training (*0.97683* for the initial model and *0.97508* for the simplified one) and validation (*0.81718* for the initial model and *0.82569* for the simplified one) is reasonably high. It should be taken into account that these are *real* data and the prediction is for *20* steps (min) ahead. Time necessary for calculations at each step is less than 1 second on a standard PC with 550 MHz and 128 MB RAM.

Predictions of the control signal compared to the measured data are depicted in the Fig. 8.34 for training data and in the Fig. 8.35 for validation data. The loss of precision of the model that this simplification results in is negligible.

The root mean squared prediction error of the initial and simplified models on the training data was *0.030* and *0.031* respectively. Application to the validation data set yields errors of *0.095* for both models. The correlation coefficients reflect similar results.

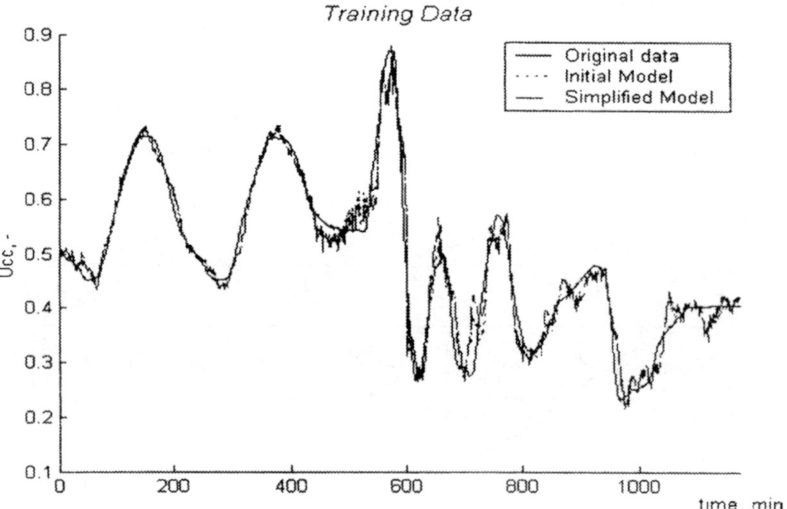

Fig. 8.34. Control signal prediction: training data

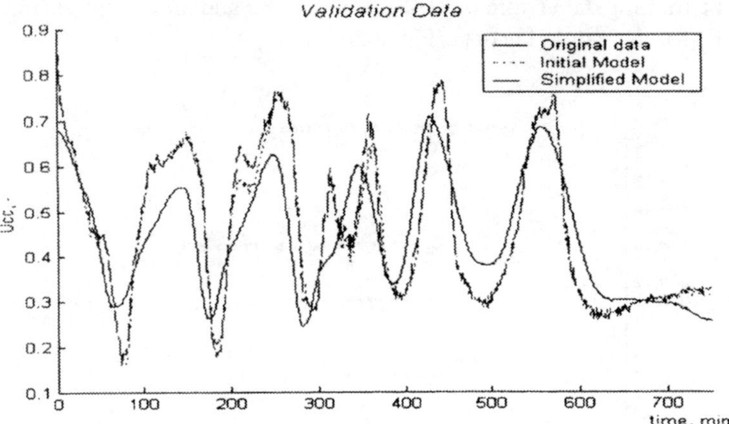

Fig. 8.35 Control signal prediction: validation data

The simplified model is found non-iteratively and has a number of advantages;

- ✓ it is more transparent;
- ✓ the linguistic terms are simpler;
- ✓ lower memory requirements result in lower computational demands and hence a faster execution time.

8.6 Refinement of Linguistic Terms' Parameters

The FRB model automatically generated in *on-line* mode could, optionally, be optimised in the sense of tuning shapes and parameters of the membership functions, including their spreads. Centres of the membership functions could also be tuned, but the range of possible variation for them is limited (to half the distance between two neighbour centres) in order to preserve the linguistic meaning of the model.

This process was described in sections 5.4.2 and 7.7.2. Essentially, it is a non-linear optimisation problem, which leads to a global (or close to it) solution. It is, however, time consuming and, to some extent, the price for improving the precision of the model is the lower level of generalisation it provides and, possibly, lower level of transparency.

Applying GA, the FRB model described in the previous section has been further optimised (tuned). After 50 generations only (parameters of the GA were $p_c=0.6$; $p_m=0.03$; $pop\_size=80$) the error of the model has been decreased more than three times for the training data (from *0.1032* K to *0.032* K) and almost twice (from *0.185* K to *0.1042* K) for the validation data (Fig. 8.36).

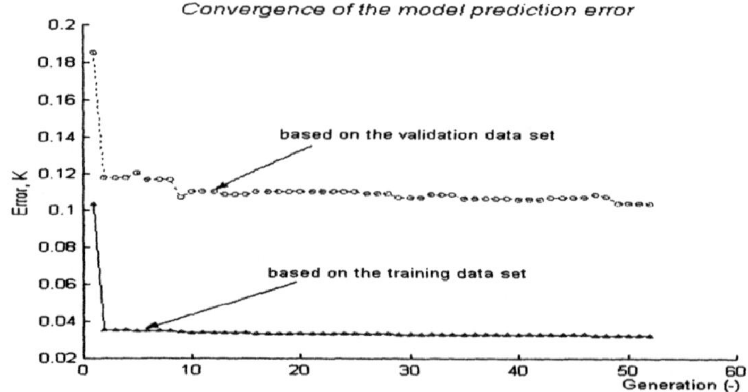

Fig. 8.36. Optional FRB model refinement by GA (convergence of the model error)

Membership functions, which describe linguistic variables, has been modified (Figs. 8.31c-8.33c).

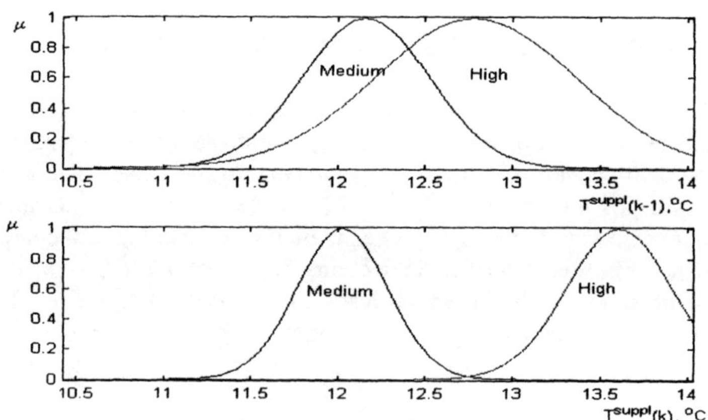

Fig. 8.31c. Linguistic terms of the supply air temperature after the refinement

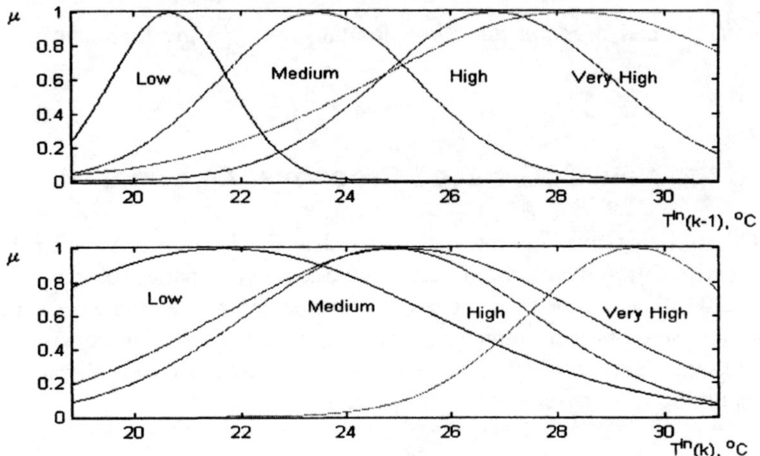

Fig. 8.32c Linguistic terms of the inlet air temperature after the refinement

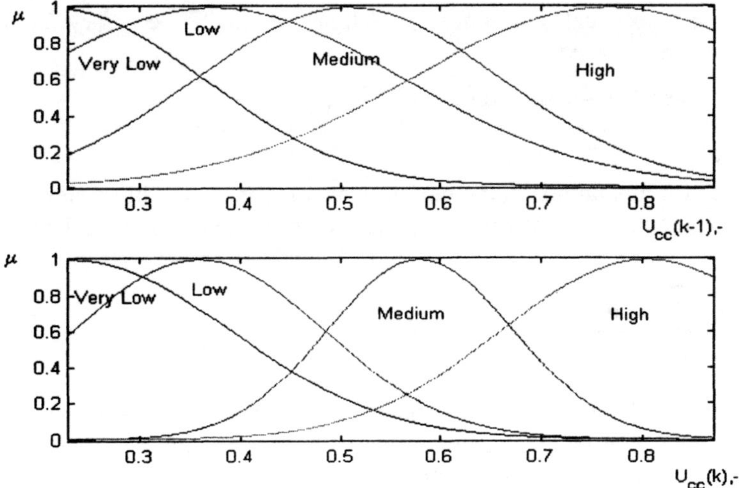

Fig. 8.33c. Linguistic terms of the control signal after the refinement

As it is seen (especially from Fig. 8.32c for the linguistic terms *High* and *Very High* for $T^{in}\big|_{k}$ and *Medium* and *High* for the $T^{in}\big|_{k-1}$) the linguistic transparency suffers.

8.7 Testing the New CoG-based Crossover Operator

The CoG-based crossover operator, considered in the Section 3, Chapter 4 has been tested first with different commonly used numerical benchmark problems (Wright and Angelov, 2000). Each test has been performed with the same GA parameters (probabilities of crossover, mutation, and population size). For consistency of the results, the same random number sequence has also been used in both cases (with and without applying CoG-based crossover operator).

8.7.1 Numerical Test Functions (NF1-NF5)

Two series of *30* runs has been carried out with all test functions:

- The first stop criterion was to find a value of the objective function ($\tilde{J}$) close to the theoretical optimum (J^0). Number of epochs necessary for both cases has been recorded:
  - ✓ using CoG-based crossover ($\#_{COG}$);
  - ✓ not using it ($\#_{conv}$).

  Based on this numbers, the rate of convergence, which illustrates the effect of the new operator in saving computational time, has been considered (Angelov, 2001b):

  $$converg = \frac{\#_{CoG}}{\#_{conv}}$$

  Number of floating point operations for both cases has also been registered and respective rate calculated. In addition the number of improvements ($\#^+$) has been registered as the number of epochs in which $CoG^i > F(chrom_j)$; $j=1,2,...pop\_size$; $i=1,2,...,\#_{epochs}$. This indicates the number of cases in which the improvements in the fitness function are due to the new operator. In all other cases normally used crossover and mutation leads to the fitness improvement;

➢ The second stop criterion is a pre-specified number of epochs ($\#_{epoch}^{max}$).

Objective function values and the number of improvements ($\#^{+}$) has been recorded. The rate of objective was calculated as:

$$obj = \frac{J_{CoG}}{J_{conv}}$$

It illustrates improvements in the precision.

8.7.1.1 De Jong's Function (NF 1)

The first test function is the so-called De Jong's function. It is continuos, convex and unimodal function:

$$f(x) = \sum_{i=1}^{n} x_i^2 \tag{8.10}$$

The global minimum of ($J^o = 0$) is at $x_i = 0$. The test with $n = 30$ variables and parameters of GA $p_c = 0.6$, $p_m = 0.03$, $ps = 30$ has been carried out for ***60*** different runs. A typical plot of the convergence is given in the Fig. 8.36. Average results of all experimental runs are tabulated in the Table 8.7.

From the results one can conclude that the new operator allows to find the same solution ($\tilde{J}$) for almost a quarter less epochs in average. In a sixth of epochs improvement of the fitness is due to the new operator.

Similarly, for a fixed number (*3000*) of epochs the result is closer to the global minimum of the objective function ($J^o = 0$) - the value of the objective function is almost 2 times less in average with improvements in a sixth of epochs due to the new operator.

It should be mention that the rate of elementary floating-point operations is practically the same as the rate of convergence, which means that the additional computational effort due to the application of the new operator is negligible.

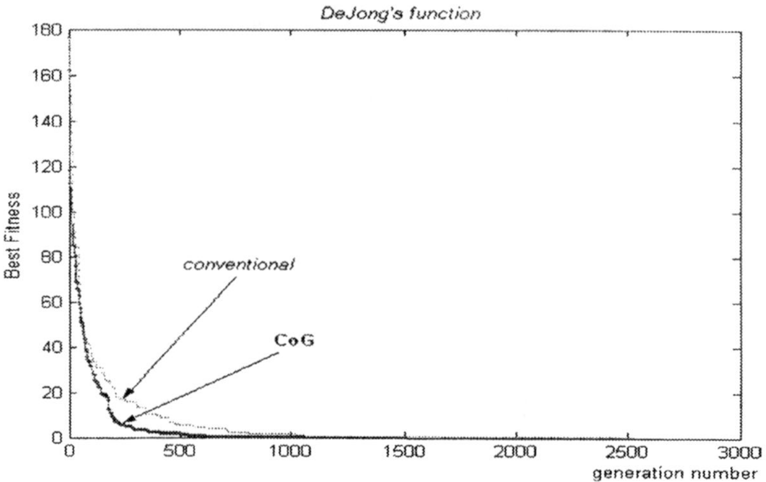

Fig. 8.37. A typical convergence in DeJong's function minimisation

8.7.1.2
Rastrigin's Function (NF 2)

Rastrigin's function has many local minima as it uses cosine modulation. Although, this function is highly multi-modal, the minima are regularly distributed.

$$f(x) = 10n + \sum_{i=1}^{n} (x_i^2 - 10\cos(2\pi x)) \tag{8.11}$$

The global minimum of $J^o = 0$ is at $x_i = 0$. GA parameters used were $p_m = 0.005$, $p_c = 0.8$, number of *bits = 10, population size =30*. A typical convergence is shown in the Fig. 8.37.

Similar conclusions could be made:

- ✓ *3* times faster convergence take place to reach $\tilde{J}$ with practically no additional computation effort;
- ✓ in a fifth of cases improvement is due to the new operator;
- ✓ *1.5* times better result is achieved for the fixed number (*10000*) of epochs;
- ✓ improvements occur in the similar proportion (*22%*) of epochs.

8.7.1.3
Sum of Different Powers (NF 3)

The sum of different powers is usually used in uni-modal tests:

$$f_i(x)=\sum_{i=1}^{n}|x|_i^{i+1} \tag{8.12}$$

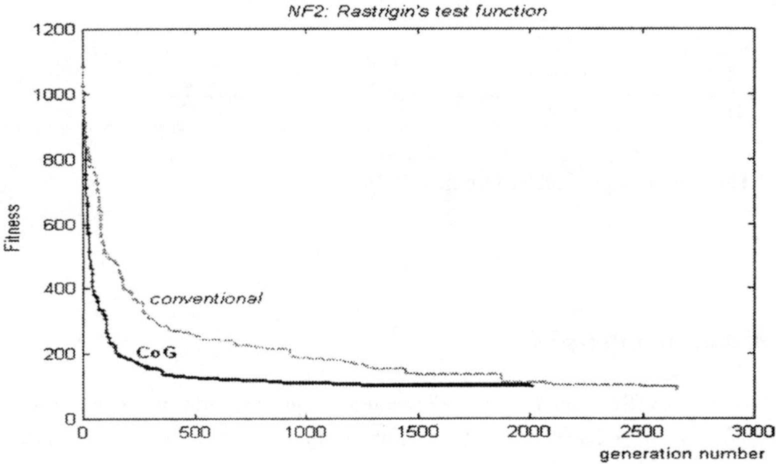

Fig. 8.38. Rastrigin's function minimisation

The global minimum of $J^o = 0$ is at $x_i=0$. A typical convergence for n=30 variables and the following GA parameters ($p_m=0.01$, $p_c=0.6$, $ps=60$, $bits=10$) is depicted in the Fig. 8.38. In this test the advantage of the new operator is even more obvious:

- ✓ $\tilde{J}$ is reached for less than *20* epochs in average when CoG operator is used;
- ✓ more than *2700* epochs are necessary for the case when it is not used;
- ✓ in each forth epoch an improvement occur due to the application of the new operator;
- ✓ *14* billion times (!!!) better result is achieved for the same fixed number of epochs (*3000*).

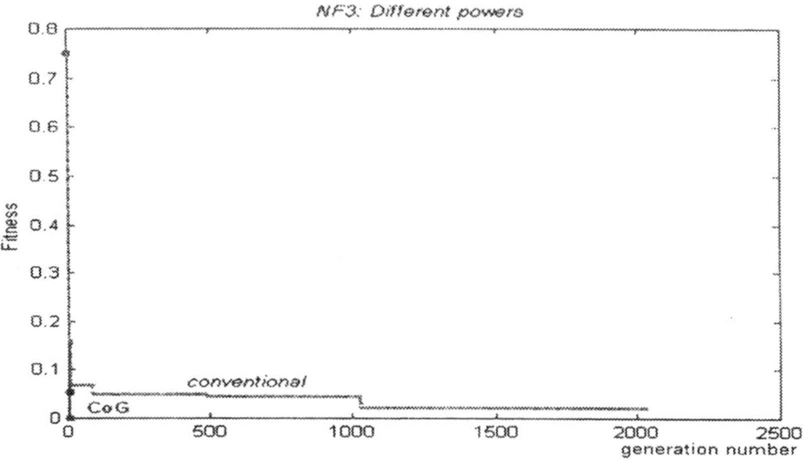

Fig. 8.39. Different powers function minimisation

8.7.1.4
Schwefel's Function (NF 4)

Schwefel's function (Schwefel, 1981) determines a geometrically distant minimum (at x_i = 420.9687; J^o = -418.9829*n) from the next best local minima. Therefore, the search algorithms often converge in a wrong direction.

$$f(x) = 10n + \sum_{i=1}^{n} x_i - \sin(\sqrt{|x_i|}) \quad (8.13)$$

A typical convergence for the test results for $n=20$ variables and $p_m=0.01$, $p_c=0.6$, $ps=30$, $bits=10$ is given in the Fig. 8.39. The results are quite obvious:

- ✓ the GA, which does not use the new operator, much more often falls into a local extreme;
- ✓ In average 5 times less epochs are necessary to reach $\tilde{J}$;
- ✓ The value of the objective function *50%* better in average is reached for the same fixed number of epochs (*1000*);
- ✓ Improvements occur in almost *90%* of epochs (!) due to the new operator.

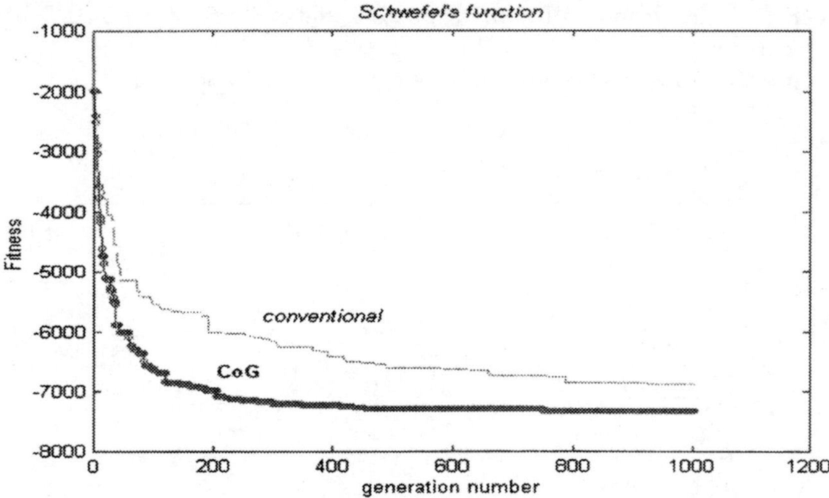

Fig. 8.40. Schwefel's function minimisation

It is interesting to mention that the standard deviation is 5 times higher for the case, when the new operator is not used, than, when it is used. This could be explained by the fact that in the case, when it is not used, the algorithm relatively often falls into local extremes.

8.7.1.5
Griewangk's Function (NF 5)

Griewangk's function is similar to the Rastrigin's function and has many regularly distributed local minimuma which are spread over the search space. The global one is at $x_i=0$; $J^0=0$.

$$f(x) = \sum_{i=1}^{n} \frac{x_i^2}{4000} - \prod_{i=1}^{n} \cos\left(\frac{x_i}{\sqrt{i}}\right) + 1 \qquad (8.14)$$

A typical convergence for the test results for $n=30$, $p_m=0.01$, $p_c=0.7$, $ps=30$, $bits=20$ variables is represented in the Fig. 8.40. Again

- ✓ 2 times faster convergence with practically no additional computation effort have place;
- ✓ In a quarter of the epochs improvements are due to the application of the new operator;

- ✓ For a fixed number of epochs (*3000*) *2* times better result (in average) is registered;
- ✓ in a sixth of epochs improvements are due to the new operator.

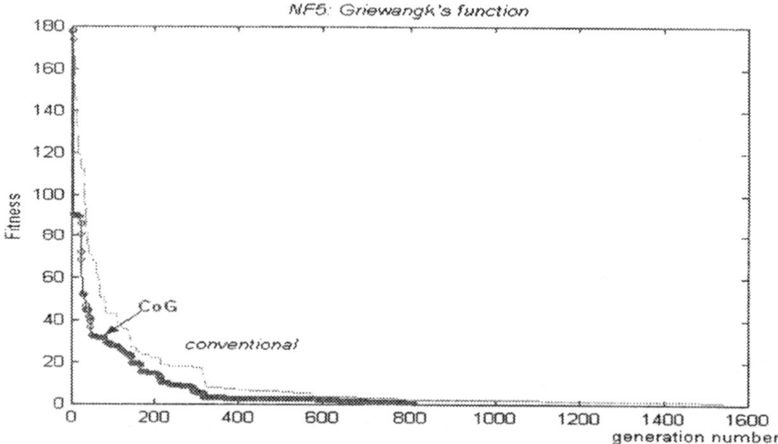

Fig. 8.41. Griewangk's function minimisation

It is interesting to mention that the standard deviation of the results of the *30* runs with the new operator are more than 3 times smaller (*0.019* instead of *0.066*) which means that the role of mutation and the randomness is less important in this case than when it is not used.

8.7.2 Optimal Scheduling of a Hollow Core Ventilated Slabs (AC)

A practical optimisation problem has been also used to test the new CoG operator (Angelov and Wright, 2000). It is to minimise the energy costs in a hollow core ventilated slab system used as a thermal storage during the night and off-peak electricity tariff periods such that not to compromise the comfort of the occupants (Ren, 1997).

The results are depicted in the Figs. 8.41- 8.44 (all other parameters, including random generator are the same in both cases).

CoG-based crossover improves significantly the convergence (Fig. 8.41) as well as the final result:

- ✓ the thermal comfort is practically not changed, while
- ✓ more effective (with *6%*) solution is achieved (normalised value of the costs is *0.90886* instead of *0.95622*).

The optimal profiles of the fan power, supply air temperature and flow rate are given in the Figs. 8.42 - 8.44 respectively.

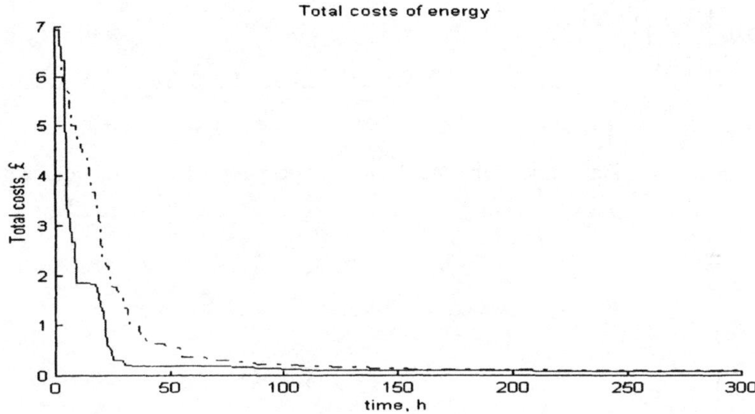

Fig. 8.42. Convergence in minimisation of the total costs of energy used (solid line represents the case when CoG-based operator is used; dash-dotted line - when it is not used)

The effect is achieved by lowering the supply air flow rate during the morning pre-cooling (Fig. 8.44) while fan is switched on an hour earlier (Fig. 8.42) with lower power. Supply air-temperature during the morning pre-cooling and after working evening hours is lower (Fig. 8.43).

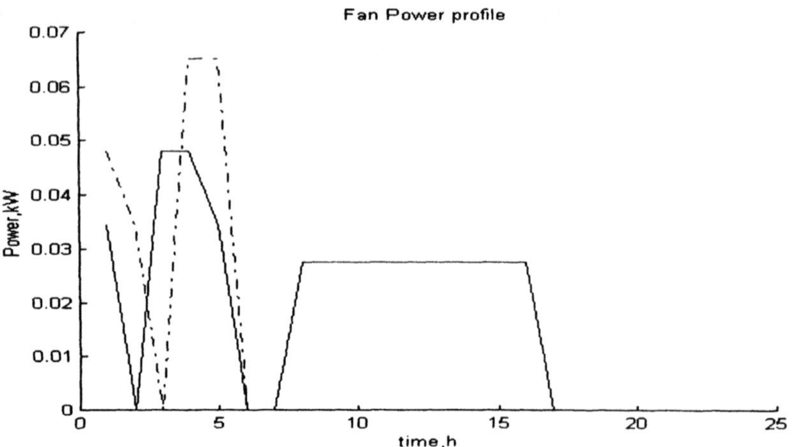

Fig. 8.43. Fan Power (solid line - with the CoG-based operator; dash-dotted line - without)

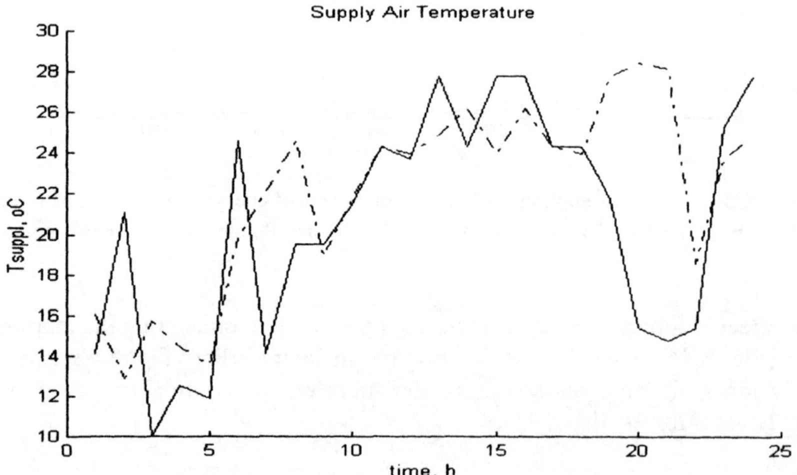

Fig. 8.44. Supply Air Temperature (solid line - with the CoG-based operator; dash-dotted line - without)

All final results are tabulated in the Table 8.7 with GA parameters used. They illustrate efficiency of the proposed new CoG-based supplementary crossover operator.

Table 8.7. Improvements effect (summarising averages from all tests)

Test functions	**Improvements effect**				GA parameters		
	converg	*#*$^+$	*obj*	*#*$^+$	p_m	p_c	ps
Test NF1	**0.753**	**16%**	**0.568**	**14%**	0.01	0.6	30
Test NF2	**0.371**	**22%**	**0.645**	**22%**	0.005	0.8	30
Test NF3	**0.006**	**22%**	**$7*10^{-11}$**	**29%**	0.01	0.6	60
Test NF4	**0.189**	**88%**	**1.563**	**93%**	0.01	0.6	60
Test NF5	**0.552**	**24%**	**0.482**	**17%**	0.01	0.7	30
Test AC	**$Cost_{CoG}$=0.90896**		**$Cost_{conv}$=0.95622**		0.001	1	80

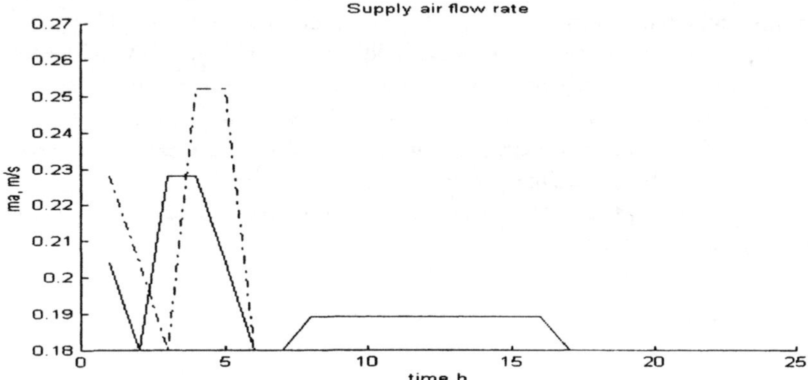

Fig. 8.45. Supply Air Flow Rate (solid line - with the CoG-based operator; dash-dotted line - without)

8.8
ICC System - Open or Closed Loop? A System Approach

The specific of the ICC systems control, by differ from most other technical systems, is the fact that human and its comfort are in the centre of the problem and, as a result, control objectives as well as some constraints are highly subjective, ill defined and variable. The main goal of an ICC is to provide the occupants of a building with comfortable and productive environment, which satisfies their physiological need.

At the same time, this should be done in the most economical way possible. The last objective became increasingly important nowadays, when ecological (the ozone hole, gas emissions, noise pollution etc.) and economical (again surging petrol prices, cost effectiveness leading to higher competitiveness etc.) concurrent requirements are pressing and have to be balanced with the comfort needs.

Temperature and relative humidity are the two basic factors to meet the psychological requirements together with the air movements and air purity. Temperatures above and below the comfort range leads to disruption of the person's metabolic processes and disturbance of his activities. The normal practice is to keep these basic physical variables within the pre-specified range, determined conservatively by the so called PPD and PMV, both of which being a statistical measure. For the temperature, for example, the recommended comfort range is 22-26°C. For the relative humidity it is 30-60% (Taylor, 1995).

The requirement depends on the type of activity undertaken (lower temperature and humidity required when the activity is higher), existing equipment, radiation etc. More importantly, the physiological requirements are *individual* and could vary. One specific occupant could prefer higher temperatures than other could, and similarly, for the humidity, light and other parameters of the comfort environment.

Current practice, however, is to design and operate ICC and HVAC systems, as their main element, under assumptions of standard occupants' comfort requirements expressed in abstract statistical measures, like PPD and PMV. In reality, occupants differ (their preferences are subjective, individual) and their presence in the occupied Zone and level of activity are time varying.

The problem of adaptation to *real* occupants' requirements and behaviour has been reported very scarcely and recently.

Its consideration could lead to a twofold improvement:

- ✓ more customised, specific user-oriented HVAC solutions;
- ✓ possible energy savings.

Still unused potential for energy savings exist in finding and keeping more economical set-points or regimes, which could be acceptable for *real* occupants, although violating abstract statistical-based requirements.

Highly desirable would be an HVAC system to be *intelligent* in the sense *to be able to learn and to adapt* to the real occupants' behaviour (Fig. 6.2). Feedback from the occupants could be:

- *Active;* reaction of the *real* occupants to change
  - Temperature set point
  - supply fan speed set point
  - possibly, supply airflow set point
  - level of lightening
  - level of shading
- *Passive* (detection by infrared sensors of the level of presence and movements of the *real* occupants).

An interesting alternative is to build a *flexible* rule-based model of the occupants' comfort requirements (expressed in the change of temperature, relative humidity, lightening and shading level) as a function of the present and past values of these variables plus the detected *active* and *passive* reactions. This model could evolve in time and would adapt to *real* occupants' behaviour (reaction) guiding the correction of the default control of the HVAC (Fig. 6.2).

Such *intelligent* HVAC could be optimised taking into account this rule-based model. Methodologically, the algorithm could be based on *e*volving **R**ule-based models, introduced in the Chapter 7.

The concept of *intelligent* ICC and *intelligent* building (***i*B**) in general has been developed and now three generations of ***i*B** have been considered. The first comprising of independent self-regulating sub-systems, second including also a network and being able to be controlled remotely, while the third has in addition the ability ***to learn*** and *continuously* ***to adapt*** *to customer requirements and the environment.*

The industrial practice is requiring new solutions and the interest to such *intelligent* systems is strong. For example, on the last 52nd Air-conditioning, Heating and Refrigerating Exposition (held in Dallas, Texas, USA in March 2000) one of the accents was the improvement of the energy management and ***customised*** comfort satisfaction (http://ashraejournal.org/features/feature6.asp). This was achieved by integration of controls for energy, lightning and security with a network, and the ability to access information in *real-time.*

Adaptation to *real* occupants' behaviour and comfort requirements using **eR** models could be considered as a third generation ***i*B** closing the loop trough the occupant/consumer.

In fact, currently existing HVAC systems are closed-loop systems, but in respect to the technical systems only. If consider the occupant/consumer as a part of the system (which have to be done because the occupant determines the inherently *flexible, fuzzy* and time-varying goal of the system) then the current schemes are open-loop systems (Fig 6.2).

In order to be a closed-loop (really controlled) system the model of the occupant(s)' behaviour, reactions and preferences is required (Fig 6.2). ***e*R** models are seen as a tool for designing and developing systems able to solve such problems.

8.9 Conclusion

Indoor Climate Control systems and particularly HVAC systems as their main energy consuming part has been considered in this chapter from the point of view of their modelling by *flexible* rule-based models. Problems of modelling and simulation of different components of HVAC systems (heating/cooling coils, fans, boilers) and sub-systems (fan-coil units) has been considered.

FRB, and especially ***e*R** models, has been proved to be an effective alternative to both first principle and *black-box* models still widely applied in practice. Their superiority in computational efforts and precision has been demonstrated on the example of *real* data.

Hybrid approach to modelling the fuel consumption of an industrial small-scale boiler has been presented, which combines FRB and first principle models. This is a promising area of application of FRB models in real engineering practice.

Modelling the thermal load of a building has also been presented. This problem complements the ICC system modelling and analysis in the process of building design. All the results illustrate high level of precision and easy of use of the FRB and ***e*R** models.

So called VL strategy, when a very limited amount of initial information is needed to start the *on-line **learning trough experience*** of the *flexible* rule-based model, has been illustrated with the same *real* data. This specifics of ***e*R** models could be very useful for robotics and control, when the whole model (controller) structure is developed trough the process of observation of the plant (object of modelling and/or control).

Dynamical signals of the position of the valve regulating the water mass flow rate trough the coil has been modelled *on-line*. The model generated has been simplified based on the similarity of membership functions (effectively their centres as they have the same form and spreads) in *on-line* mode as described in sections 5.4.1 and 7.7.1. The simplified model has two times less membership functions, while providing the same precision both in training and validation.

As a separate test, the linguistic terms' shape (membership functions parameters, including their spreads) has been modified using GA. This is an iterative process and, thus, a time consuming one. Therefore, it is optional. For relatively small number of generations (*50*) significant improvement of the precision (*3* times in training and *2* times in validation) has been achieved. As it has been mentioned, however, the price is paid by the lower linguistic transparency of the resulting FRB model.

The new (centre-of-gravity-based) crossover operator, introduced in Section 4.3 has been tested by a number of numerical functions and on a *real* problem of optimal scheduling of hollow core ventilated slabs. It proves to be significantly more efficient and has been used when GA has been applied.

Finally, the problem of ICC systems has been treated from the point of view of the system theory. It has been mentioned that the loop of control of such systems, which inherently incorporate humans, should be made in respect to the consumer.

Current practice ignores this problem, and therefore the *fuzzy, vague* goal of such systems (*to satisfy comfort requirements of the occupant(s)*) has normally be substituted by a simplifications such as standard requirements, ignoring the individuality, time-variations and complexity of *real* problems. As a result, poor performance and, sometimes, dissatisfaction of the use of such systems has been reported.

9 *ON-LINE* MODELLING OF FERMENTATION PROCESSES

9.1 Bio-processes - Specifics of their Modelling

Biotechnological processes (an important part of which is fermentation processes) are characterised by:

- ✓ Uncertainties;
- ✓ non-stationary dynamics;
- ✓ non-reproducibility (uniqueness);
- ✓ existing of non-quantified factors, like
  - *smell;*
  - *taste;*
  - *morpho-physiological specifics;*
  - *colour* etc.

Practically, subjective estimations of experienced (bio)technologists are crucial for a good performance of any biotech process (Shimizu and Ye, 1995; Linko, 1988).

The basic difficulties in modelling and control of biotechnological process (which are used in foodand pharmaceutical industries, medicine and waste water treatment etc.) are (Staniskis et al., 1988):

✓ **Lack of knowledge and information** about the process.
The level of knowledge for most of complex transformations, related with metabolism of micro-organisms is not enough. Each cell can be considered as a separate multi-component, self-organising, multi-factor system, which is in permanent energetic, and metabolite exchange with the environment. This hampers the prediction of process reaction to the control actions and therefore the design of effective control algorithms.

✓ **Limited possibilities for interpretation** of the information received especially the **subjective and qualitative** one.
It often includes unclear, *fuzzy* inherent purpose of fermentation processes itself in the so-called secondary metabolite synthesis (for example, antibiotic and vitamin production) there is no direct relation between the product synthesis and the primary metabolism.

- ✓ **Lack of sensors and analysers** for permanent control of the state of the process.

  There exist sensors for measurement of physical and chemical parameters like temperature, pH, rH, dissolved oxygen concentration, r_{CO_2}, but rarely exist precise automatic sensors for the most important parameters: concentrations of the cell mass, substrates and products of metabolism;

- ✓ **Uniqueness.**

Each run of fermentation is different by the type of used tanks, species, feeding type extracting product etc. From formal point of view, changing the strain, tank or inoculate, the process is yet a new one and a new model and control system is necessary. The reason for this is that the object of control (micro-organisms) is a live creature.

First principle models of fermentation processes are based on mass- and energy balance and are represented as sets of differential equations (2.5)-(2.7). As it has been mentioned in the Chapter 2, this expression is computationally very expensive, especially for optimisation and control, and, therefore, many assumptions and simplification has been made to simplify it.

For example, in an optimal control problem based on such a model transversally (boundary) condition has to be defined. Normally it is defined as

"concentration of the feed substratum at the end of the batch fermentation to be zero":

$$S_N=0 \tag{9.1}$$

In reality, it is an idealisation, as substratum concentration never could reach this value (Angelov and Tzonkov, 1993). In this case more appropriate is the following *flexible* set:

"*S_N to be* ***as much as possible close to*** *zero*":

$$S_N \cong 0 \tag{9.2}$$

where $\cong$ denotes *fuzzy* equality.

It could have a membership function of the following type:

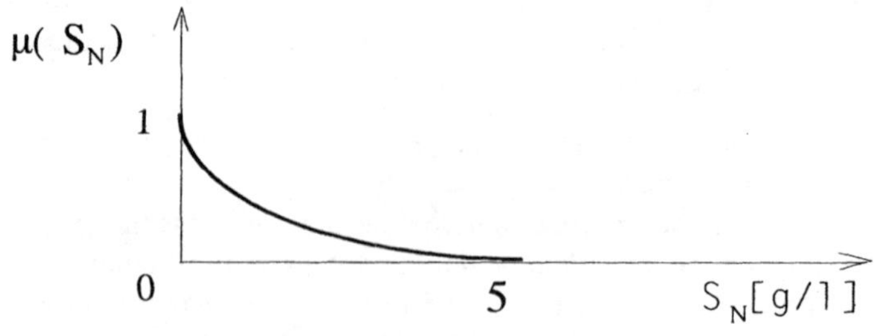

Fig.9.1. Flexible transversally condition 'Substrate to be **as much as possible close to** zero'

$$\mu(S_N) = \begin{cases} \exp(-S_N); & S_N > 0 \\ 0 & : S_N < 0 \end{cases} \tag{9.3}$$

The meaning is that:

- although the most desirable ($\mu = 1$) is the zero concentration at the end of the process ($S_N = 0$);
- low concentrations are also acceptable;
- the lower the concentration is the higher is the degree of its acceptance.

9.2 *e*R Model of a Fermentation Process

An alternative for modelling such complicated processes represents FRB models. They have been applied already to this problem (Linko, 1988; Shimizu and Ye, 1995). The structure and parameters of these models, however, did not change, while the low reproducibility of fermentation runs leads normally to a poor performance of any type of models applied to the next run.

***e*R** models, introduced in the Chapter 7, are able to adapt their structure as well as parameters in *on-line* model in respect to any new data sample. It is a well-known difficulty in this type of processes to have measurements of concentrations of spices in relatively distant moments only (normally not less than an hour). *Real-time* measurements are possible sometimes mainly for gas concentrations, temperatures, and *pH*. Recently, approaches for *on-line* measurement (optically based) of cell mass concentration has also been developed, but they are far not widely used.

9.2.1 *Lactose* Oxidation - Process Specifics

A fed-batch fermentation of lactose oxidation from natural substratum in fermentation of *Kluyveromyces marxianus var. lactis MC 5* is considered as an example (Angelov et. al., 1996). The model includes the dependence between concentrations of the basic energetic substrates: lactose (S) and dissolved oxygen concentration (DO).

This process is not well studied. There does not exist a general mathematical model of the microbial synthesis because of the extreme complexity and great variety of living activity of the micro-organisms, although various models of different parts of the whey fermentation exist. *Flexible* rule-based, and particularly ***e*R**, models are considered as an alternative technique, which is very effective in cases of complex and sophisticated plants.

Six runs of fermentation where carried out in aerobic batch cultivation of *Kluyveromyces lactis*. A laboratory bio-reactor ABR 02M with capacity 2 *l* has been used. The strain *Kluyveromyces marxianus var. lactis MC5* has been cultivated under the following conditions (Angelov et. al., 1996):

✓ Nutrient medium with basic component - whey ultra filtrate with lactose concentration 44 *g/l* has been used. The ultra filtrate has been derived from whey separated in production of white cheese and de-proteinisation by ultra filtration under the following condition:

- $T = 40\text{-}43\,°C$;
- input pressure $p^{in} = 650\ Pa$;
- output pressure $p^{out} = 600\ Pa$.

The ultra filtrate has been used in native condition with lactose concentration 44 *g/l*. Nutrient medium consist of:

- *$(NH_3)HPO$ - 0.6%*;
- yeast's autolisate - *5%*;
- yeast's extract *-1%*;

Slightly acidic medium has been composed ($pH \in [5.0\,;\,5.2]$).

✓ The velocity of the airflow has been *1.0* $\dfrac{l}{l\min}$ up to the *4*th hour and *2.0* $\dfrac{l}{l\min}$ up to the end of the process under continuos mixing $n=800\ rpm$;

✓ Temperature has been *29 °C*;

The following changes of the microbiological process (lactose conversion in yeast's cells to protein) have been observed and studied during the strain growth (Angelov et. al., 1996):

✓ lactose concentration in fermentation medium in oxidation and assimilation of lactose by *Kluyweromyces lactis* has been determined by enzyme methods using UV tests;

✓ concentration of the cell mass and protein contents has been determined on the basis of the nitrogenous contents;

✓ concentration of the dissolved oxygen in the fermentation medium in the process of oxidation and assimilation of lactose has been determined by a laboratory oxygen sensor.

9.2.2 Experimental Data

Experimental data for the cell mass, substrate and oxygen concentrations are tabulated in the Table 9.1. The data has been taken at intervals of 2 hours for all six runs

Run No	time, h	0	2	4	6	8	10	12	14
1	X, g/l	0.2	0.57	2.91	5.9	11.6	17	21.8	
	S, g/l	44	41	32.5	20.1	13.5	5.8	0	
	DO, mg/l	6.57	5.68	3.7	0.97	0.15	0.18	0.15	
2	X, g/l	0.2	0.48	3	6.2	11.6	17.5	22.0	23.8
	S, g/l	44	41.8	39.3	22.8	11.5	6.1	1.8	0.3
	DO, mg/l	6.7	5.82	3.98	1.01	0.18	0.18	0.28	0.18
3	X, g/l	0.2	0.52	2.86	5.09	11	16.8	22.5	23.6
	S, g/l	44	41.3	31.5	21	10.1	7.7	4.3	2.1
	DO, mg/l	6.6	6.16	4.71	2.21	0.25	0.18	0.69	0.60
4	X, g/l	0.2	0.46	2.58	5.8	10.9	18.0	21.9	
	S, g/l	44	40.7	32.4	24.1	18.8	10.5	3.0	
	DO, mg/l	6.78	5.54	4.04	0.71	0.14	0.18	1.02	
5	X, g/l	0.2	0.7	3.15	6	11.6	20.1	20.1	22.4
	S, g/l	44	41.9	36.8	21.4	14	0.1	0	0
	DO, mg/l	6.99	5.93	3.45	2.8	0.5	0.09	0.09	1.92
6	X, g/l	0.2	0.58	3.2	6.5	11.4	15.8	19	21.2
	S, g/l	44	41.3	35	24	15.1	7.8	0.03	0
	DO, mg/l	6.26	5.59	2.54	0.13	0.07	0.07	0.28	1.65

Table 9.1. Experimental data from 6 fermentation runs (Angelov et. al., 1996)

9.2.3 Modelling the Process

9.2.3.1 First Principles-based Model

The dynamic of biotechnological processes can be described by first principles based model using differential equations which represents mutual dependencies of cell mass (*X*), substrate (*S*) and dissolved oxygen concentrations (*DO*) and agitation rate (*n*), temperature (*T*), *pH* etc.

Discretisation is usually possible due to the slow changes in these processes. A general discretised model of a biotechnological process could be given as:

$$X_{k+1} = \mu_X (X_k, S_k, DO_k) \qquad k=0,...,N-1 \tag{9.3}$$

$$S_{k+1} = q_S (X_k, S_k, DO_k) \tag{9.4}$$

$$DO_{k+1} = q_{DO} (X_k, S_k, DO_k) \tag{9.5}$$

where μ_X , q_S, q_{DO} denote non-linear specific rates

The main difficulty in building such a model is to identify the highly non-linear and time-varying specific rates μ, q_S, q_{DO}.

9.2.3.2 eR Model

As it could be seen from the experimental data (Table 9.1), however, the reproducibility is low:

- ✓ fermentation runs *NoNo1 and 4* ends up earlier (in fermentation *No4*, additionally, as in fermentation runs *NoNo2 and 3* the feed substrate - lactose - is not fully utilised although the cell growth stopped for some reason);
- ✓ in runs *NoNo4, 5* and *6* the dissolved oxygen concentration at the end of fermentation starts to increase slightly in discordance to the other fermentation runs.

This is a typical case with this type of processes. This is the reason first principle models did not give good results in their modelling. Additionally, it is extremely difficult to accommodate all factors, which influence the cell mass growth into the model (stress, cells age, physiological state etc.).

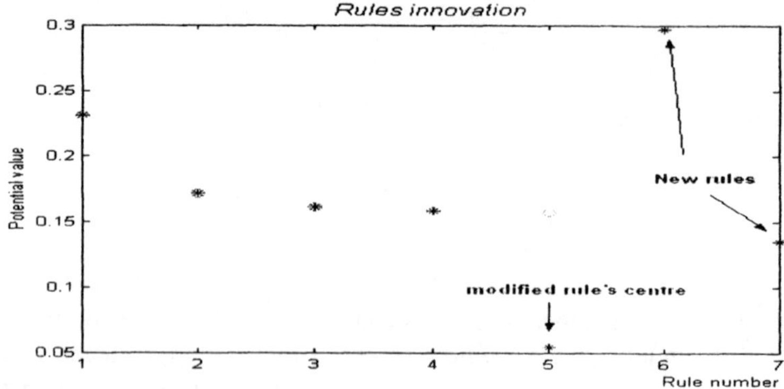

Fig 9.2. **eR** model structure innovation in *on-line* mode

Flexible rule-based and, particularly ***e*R** models, are a promising alternative. VL strategy has been applied, because the experimental data has been collected relatively rarely (every 2 hours), which is another typical hurdle in biotech process modelling (Staniskis et. al., 1988). ***e*R** model has been able to start with *6* (six) data points available from the first experimental run only and to generate successfully its structure (Fig. 9.2) and parameters (Fig 9.3) in *on-line* mode (step by step).

This process has been driven by the informative potentials up-date in *on-line* mode as described in the Chapter 7. It is graphically represented in the Fig. 9.4 and Fig. 9.5. From the second figure it is seen that two new rules have been generated.

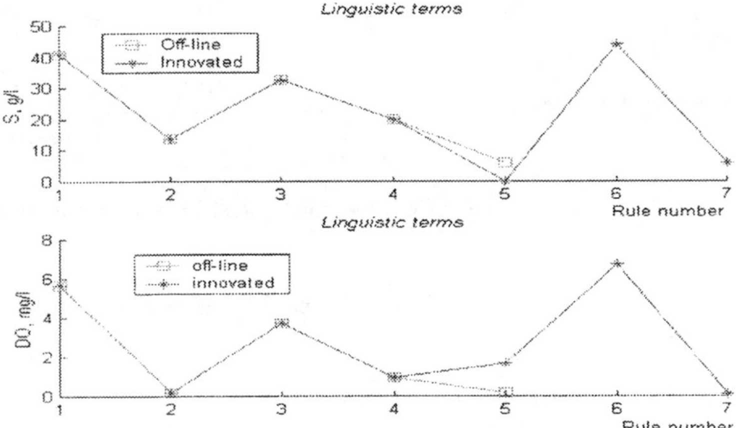

Fig 9.3. **eR** model parameters *innovation* in *on-line* mode

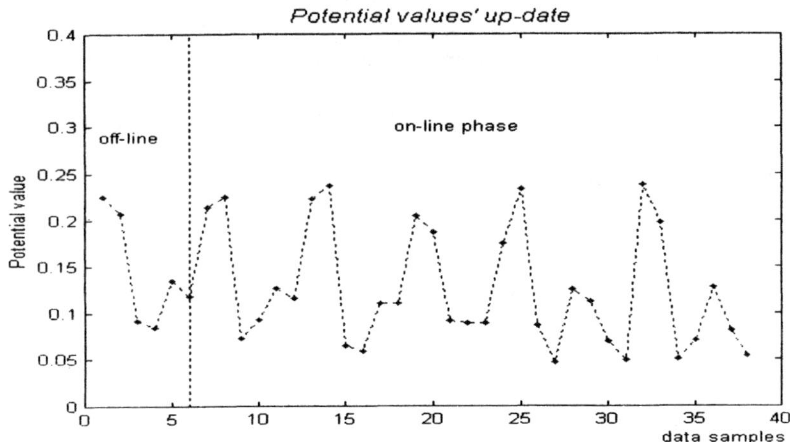

Fig 9.4. Potentials up-date

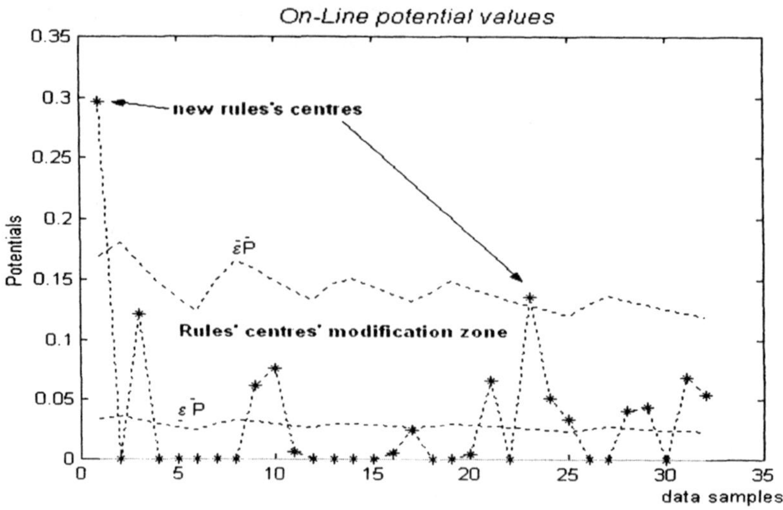

Fig 9.5. **eR** model structure *innovation* in *on-line* mode based on the potential values

One other rule has been modified (Fig. 9.2) and the final *flexible* rule-based model has the following structure:

R_1: ***IF*** *(S is Very High)* ***AND*** *(DO is Very High)*

$$\textbf{\textit{THEN}}\ X = a_0^1 + a_1^1 S + a_2^1 DO \qquad (9.6)$$

R_2: ***IF*** *(S is Very Low)* ***AND*** *(DO is Very Low)*

$$\textbf{\textit{THEN}}\ X = a_0^2 + a_1^2 S + a_2^2 DO$$

R_3: ***IF*** *(S is High)* ***AND*** *(DO is High)* $\textbf{\textit{THEN}}\ X = a_0^3 + a_1^3 S + a_2^3 DO$

R_4: ***IF*** *(S is Low)* ***AND*** *(DO is Low)* $\textbf{\textit{THEN}}\ X = a_0^4 + a_1^4 S + a_2^4 DO$

R_5: ***IF*** *(S is Very Medium)* ***AND*** *(DO is Medium)*

$$\textbf{\textit{THEN}}\ X = a_0^5 + a_1^5 S + a_2^5 DO$$

R_6: ***IF*** *(S is Extremely High)* ***AND*** *(DO is Extremely High)*

$$\textbf{\textit{THEN}}\ X = a_0^6 + a_1^6 S + a_2^6 DO$$

R_7: ***IF*** *(S is Extremely Low)* ***AND*** *(DO is Extremely Low)*

$$\textbf{\textit{THEN}}\ X = a_0^7 + a_1^7 S + a_2^7 DO\,)$$

Membership functions parameters of linguistic terms of *flexible* variables ***Substrate*** and ***Dissolved oxygen concentrations*** used in the model are represented in the Fig. 9.6.

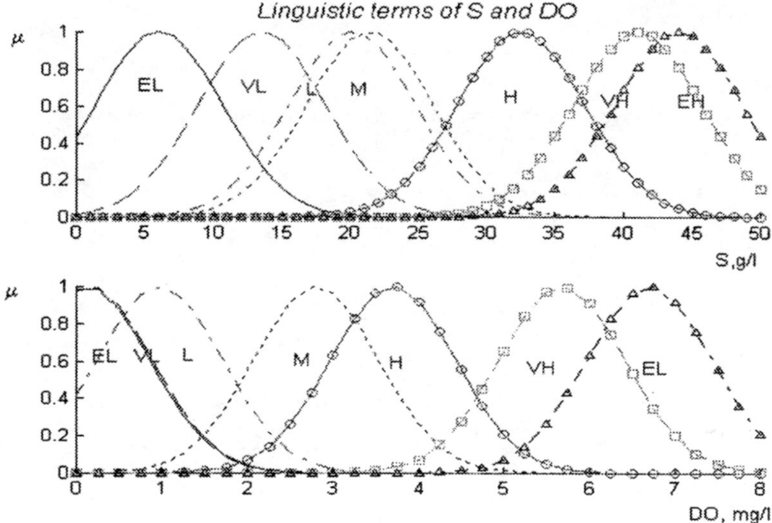

Fig 9.6. Membership functions of the linguistic terms for S and DO used

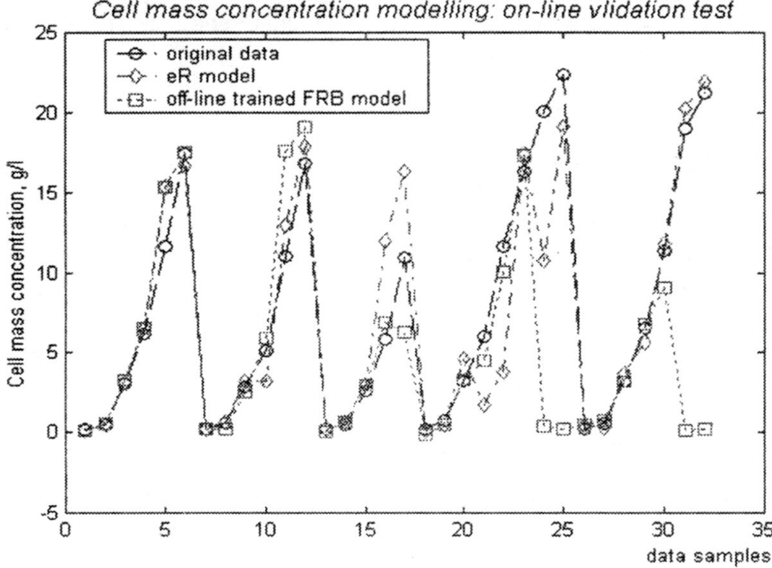

Fig 9.7. *On-line* validation test (original data, *off-line* trained FRB an **eR** model)

Performance of the ***e*R** model has been compared to the original experimental data (Table 9.1). The results of the validation test (root mean square error in prediction of the cell mass concentration *0.51843 g/l* for the ***e*R** model and *1.3188 g/l* for the *of-line* trained FRB model) illustrates the advantage of the ***e*R** models in terms of precision (Fig. 9.7 and Fig. 9.8).

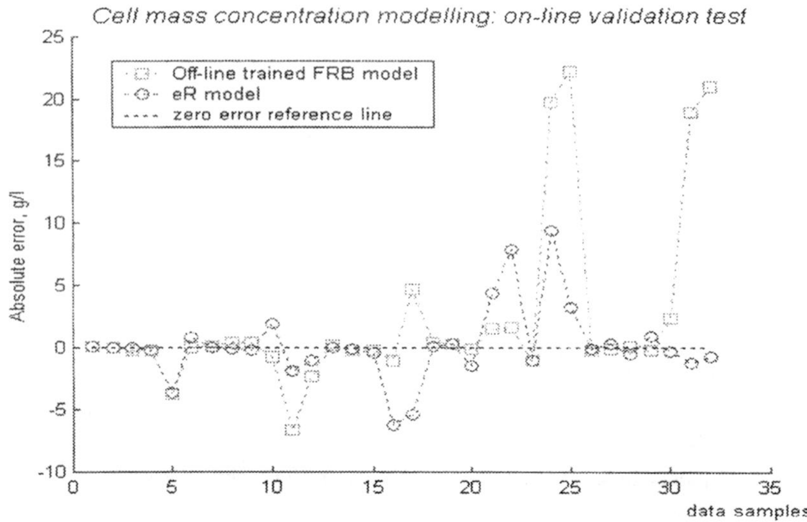

Fig 9.8. *On-line* validation test (original data, *off-line* trained FRB and **eR** model)

9.2.3.3
Analysis of the Results

It could be seen that the performance of the ***e*R** model is very good for runs *NoNo 2, 3 and 6*. Performance of the *off-line* trained FRB model is also good for the runs *NoNo2* and *3*. But, for the runs *NoNo 5* and *6* it is unsatisfactory (Figs. 9.7 and 9.8).

The performance of both models for the run No 4 is poor. One possible reason could be that this experimental fermentation run is not a *typical* one: the cell mass growth ends up earlier, while the feed substratum is not fully utilised (Table 9.1). The aggregate performance of the eR model is significantly better: almost three times lower root mean squares error in validation.

9.3
Conclusion

Application of ***e*R** models for a fed-batch fermentation process (lactose oxidation from natural substratum in fermentation of *Kluyveromyces marxianus var. lactis MC 5*) has been considered in this chapter as a demonstration of the effectiveness of these type of *flexible* yet transparent models.

It has been demonstrated that the problem of the lack of sensors and low reproducibility of these type of processes could well be tackled by ***e*R** models. They could effectively be used for *on-line* estimation of highly non-linear characteristics of the process and could develop their structure and parameters in *on-line* mode.

10 *INTELLIGENT* RISK ASSESMENT

In this chapter applications of the *non-linear* and *quasi-linear* identification approaches to FRB model design have been considered on the example of risk assessment in several areas:

- ✓ Insurance and banking (creditworthiness assessment by *evolving intelligent* decision support systems);
- ✓ *Intelligent* risk assessment in civil aviation;
- ✓ Construction industry (distributed *intelligent* system for tendering evaluation in large-scale international construction projects);

They represent previous experience of the author and some new projects in this area and should not be considered as limiting to the scope of possible applications of the approach treated in the book.

10.1 Application of *e*R Models in Creditworthiness Assessment

A soft computing-based tool for decision support has been represented in (Angelov and Lakov, 1998). It makes more convenient and *intelligent* the solution of different assessment problems, which usually arise in administration, financial and business activities like:

- creditworthiness assessment;
- insurance risk assessment;
- choice of alternatives;
- investments project estimation, etc.

thus increasing the Machine Intelligence Quotient (MIQ). The decision kernel is based on *flexible* rules, which are defined based on the quantitative and qualitative data available.

10.1.1
Creditworthiness Assessment: Problem Specifics

Normally, there exist a number of data sources in every enterprise (bank, insurance agency, consulting and industrial company, etc.) like databases containing economical results, human resource characteristics, ecological aspects of its activity, etc. A part of these data are quantitative, like:

- profit;
- sales;
- production, etc.

but the other part is of qualitative nature:

- ♦ product quality;
- ♦ human resources abilities;
- ♦ ecological problems, etc.

Nevertheless, the last group is significant for the decision making process.

There exist a great majority of *conventional* (crisp) decision support programs and tools. Most of them, however, neglect the second group of factors. Some of them allow for dealing with qualitative knowledge but they give a general framework only, which is not oriented to specific risk assessment problems.

The ability of a typical rule-based system to reason are exclusively due to the fixed set of knowledge (rules), which it is programmed to use. A typical rule-based system can express no more decision-making ability than is explicitly contained within the knowledge base. ***e*R** models, represented in the Chapter 7, offer an ability to add, modify, analyse the rules generated in *real time* due to the change in the environment or the object itself.

The considered rule-based system combines the *flexibility* (using of qualitative knowledge represented by *flexible* sets, graphic user interface, convenient data import and export, etc.) and the problem oriented definition which allows to use it without deep *a priory* knowledge of computer literacy.

10.1.2
Flexible Rule-based System

User friendly interface is realised for the input data. There are three types of input data that can be downloaded into the tool:

- ✓ digital (crisp) information from the reports and balance sheets of the companies under consideration.

They could contain raw data, which usually exist in a database of the user (bank or an insurance agency). Usually the factors are expressed in absolute values in pre-defined ranges;

- ✓ linguistic factors.
  For example: *'Low', 'Medium', and 'High'.* This concerns the vague (*flexibly*) defined variables such as human factors, which are normally defined by experts;
- ✓ statistically defined factors as a result of pre-processed samples of time series, histograms, etc.

They are expressed by values from the range [0; 1] and have probability characteristics of an uncertain information. Similarly to the previous type, there are some pre-defined basic types of these factors, which can be extended, if necessary. They are also downloaded from the initial database.

Inference Engine

The inference is based on several *typical* types of inference schemes (Mamdani- or Sugeno-type). Without loosing generality, they are defined as two input-single output basic modules, which form a hierarchical structure of the whole system. This increases the *flexibility* of the model and makes it more generic.

Additionally, it makes possible on the intermediate level to obtain an information starting from the beginning to the very final assessment, so that an expert can actively analyse and influence this process.

Output interface is represented as a set of recommendations of assessment within unit interval, percentage or as linguistic qualifiers: '*Low*', '*Moderate*', '*High*' etc. in dependence on the problem faced.

10.1.3 Credit Risk Assessment by the *Flexible* Rule-based System

The credit risk is treated as a multi-attribute problem using a number of factors. The risk assessment is calculated using *flexible* inference procedure. Usually, credit risk assessment is defined as multi-criteria decision analysis using a definite number of factors, which reflect all aspects of this bilateral process, i.e. a relation between the two partners (lender and borrower).

They can be presented in the following three groups (Angelov and Lakov, 1998):

- ✓ Estimation of the borrower's potential. It consists of the following parts:
  - volume of the available **assets** (firm, intangible assets, etc.);

- level of the borrower's **production** and its **realisation**;
- **human factor** of the borrower's manager.

✓ Estimation of lender's abilities:

- a quantity of the **net capital**;
- feasibility in the **partner's relations**.

✓ Global economic conditions:

- stability of the **economical growth**;
- **stability of the country** and international relations;
- **state debt** position.

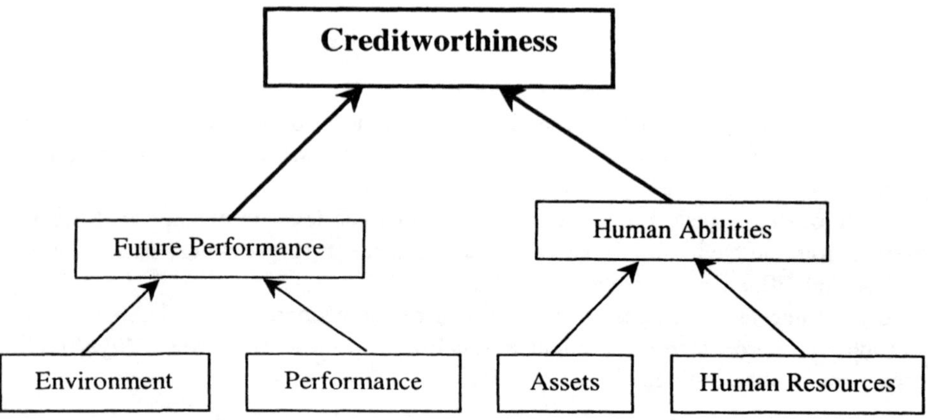

Fig.10.1. Hierarchical scheme of the C*reditworthiness* assessment

The third group of factors may be expressed in an implicit form into first two groups. Therefore, a simplified hierarchical structure is used, which reflects following levels, connecting all the representative factors under consideration (Fig. 10.1).

Some of the factors have proportional, other inversely proportional influence to the overall risk. The hierarchical structure represents the upper level of the risk of credit lending assessment. Going deeper into each category, the next, lower levels are represented as follows (Angelov and Lakov, 1998):

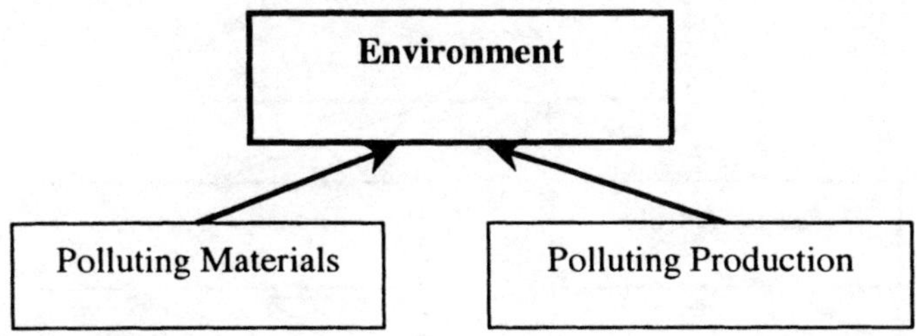

Fig.10.2. Scheme of the *Environment* estimation

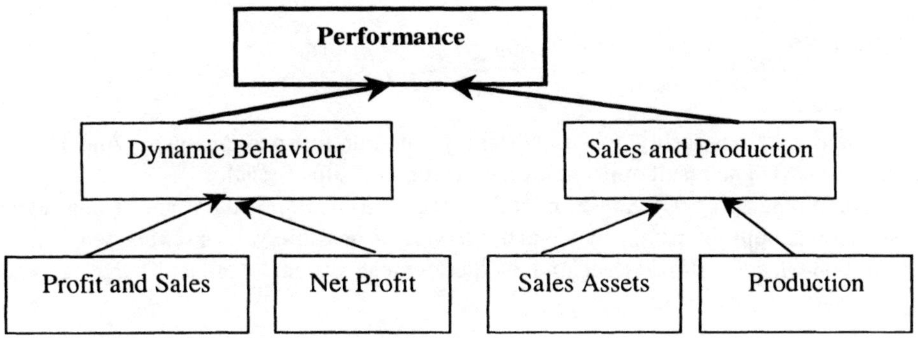

Fig.10.3 Scheme of the *Performance* evaluation

A part of the estimating factors (assets, production, sales, etc.) have crisp numerical evaluation (normally in US dollars or Euro). Others could be obtained from the statistical information based on collected representative samples (anticipated growth, rates of earning, of interest etc.). Another group of factors is subjective or *flexible* in nature (human and ecological factors, potential etc.).

The crisp approach obviously fails in obtaining good results in regard to such different parameters. On the other hand *flexible* approach could be applied even for interpretation of quantitative and statistical information.

A great variation of statistical samples points to the necessity to apply uncertainty of other type than probabilistic. Thus, a part of factors are considered to be crisp and the others are qualitative for the specific context.

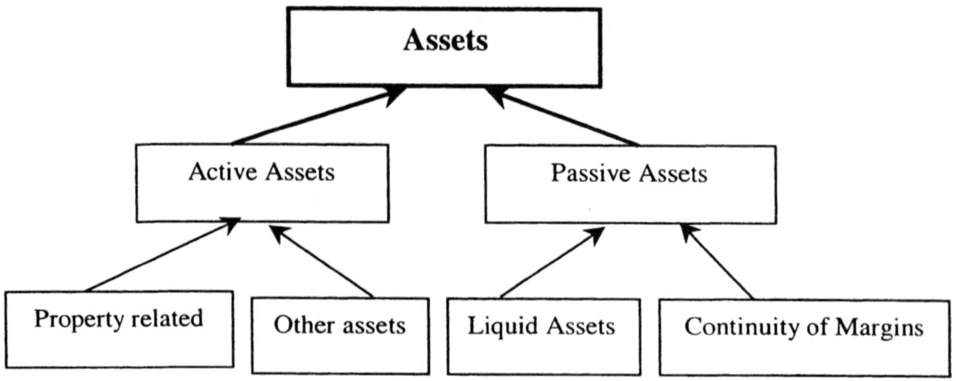

Fig.10.4. Scheme of the *Assets* estimation

There is a possibility for user-friendly intervention and analysis (Angelov and Lakov, 1998). The result appears level by level and value by value.

Combining properly these mixed (crisp and *flexible*) data and using expert knowledge about their importance and intervals of possible values (expressed in initial definition of each *flexible* set and of hierarchical system in general) the aggregate conclusion could be made.

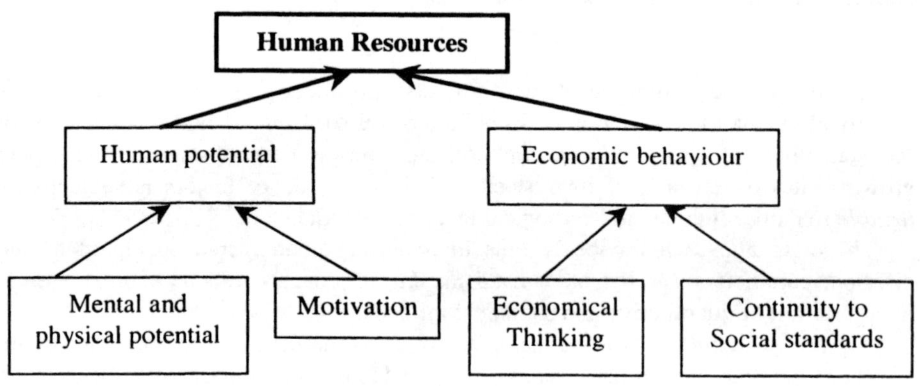

Fig.10.5. Scheme of the *Human resources* evaluation

At the end, a post-solution analysis could be done, which gives additional information about the state of the problem.

Each new set of data is considered as an additional training data sample and the system is up dating it structure, as discussed in the Chapter 7.

10.2 *Intelligent* Evolving System for Risk Assessment in Civil Aviation

10.2.1 Specifics and Importance of the Problem

Risk and safety has always been considered as the most important tasks in both planning and operating civil aviation. Recently, they have become of growing importance and concern as the sector is predicted to continue to grow on an average annual rate of at least 5-6% (Janic, 2000).

Increasing traffic densities will inevitable increase the operational complexity within the system, which would create the highly stressful working environment for the whole aviation staff. Under such circumstances, the probability of different kinds of human errors caused by overloading, which directly or indirectly may cause an air incidents and accidents, would increase.

There has been a permanent scarcity of capacity of the aviation infrastructure (airports and air traffic control system). Numerous, particularly environmental, barriers hinder further 'free' and 'unlimited' growth of the sector in terms of the use of the existing and planning of the new infrastructure.

For example, by keeping the accident rate at the same level, the number of accidents will increase in line with the growth of traffic. Some analysis have shown that under such circumstances, nearly every week a severe aircraft accident may be expected to happen somewhere in the world by the year 2015 (Janic, 2000). Due to the further globalisation of the world aviation, such development may unpredictably affect national aviation sectors.

Another process, which could affect safety, is the privatisation in the form of so-called PPP (Public-Private Partnership). This will enable separation of 'safety regulation' from the 'service provision'.

10.2.2
Intelligent Technologies for Risk Assessment in Civil Aviation

New technological challenges, such as Internet and mobile communications, *intelligent* and *agent* technologies, linguistic rule-based models etc., offer prospectus for new ways and techniques for risk assessment and prevention of incidents and accidents. They could be seen as a vehicle for gathering more data (both quantitative and qualitative), providing them in a *real time* and making possible manipulation of new types of data (linguistic, vague, uncertain, and qualitative).

Using such technologies it could be possible to successfully manage the system in a sustainable way. That means increasing the aviation output in terms of the volume of flights, number of passengers and/or passenger kilometres and constant/decreasing number of incidents and accidents (both in absolute and relative terms).

Recently, the so-called *intelligent technologies* or ***i*-tech**, and especially *fuzzy* set theory, has been intensively applied to the risk assessment (van den Brink et. al., 1995; Seo, 1999). *Flexible* rule-based (FRB) models represent a powerful tool for transparent representation of complex cause-reason type of dependencies.

They provide linguistic description in plain English and allow easy analysis by an expert in the field of safety and risk without necessity to be familiar with the complex mathematical modelling and identification techniques. They could describe quantitative as well as qualitative and imprecise, vague data, such as *Level of experience* of the crew, which could be *Very Experienced, Experienced* and *Modestly Experienced* or *Relatively Inexperienced.* The psychological characteristics of a given crewmember could be *Very Stable, Stable* or *Unstable.*

The cumulative value for the whole crew could be found by appropriate aggregation of these linguistic variables. FRB models allows aggregating of qualitative and quantitative data, such as

- Number of flight hours;
- Age of the aircraft;
- Weather data;
- Traffic density.

There is a clear need for a convenient tool, which would be able:

- ✓ to analyse and assess the risk and safety in civil aviation by the use of the historical data;
- ✓ to predict the risk and safety for future traffic scenarios by using past experience.

In particular, the tool, which would be able to handle qualitative inputs, which prevail at many files about air incidents and accidents and to give estimates of risk and safety, is currently not available. It has to be mention that practically there is no yet available result of integration of ***i*-tech** into the risk and safety assessment in civil aviation.

***e*R** models offer an opportunity to develop an *intelligent soft-computing* tool, which would allow more comprehensive analysis, assessment and prediction of the risk and safety in the civil aviation at different *levels of aggregation* under given institutional and operational circumstances. They may concern the airline, aircraft or crew characteristics, airport and air-traffic controller characteristics, related risk and safety regulation, and weather conditions considered as an external factor.

If used in combination with Internet and mobile communications it will make possible a *real-time* assessment of a given flight, suitability of a given crew or aircraft for a particular flight under specific weather conditions. In general, it will be used for analysis and planing purposes.

The stages for developing such a tool could be:

- Analysis of the available databases on the air incidents and accidents in order to classify them with respect to similarity of causes, circumstances and consequences. A thorough analysis of the cause-reason dependencies needs to be carried out.
  The framework of the hierarchical structure of the risk assessment has to be established with formulation of different groups of factors, such as *Crew, Aircraft, Traffic, Weather* etc. Their sub-structure needs to be established, such as *Age, Experience, Psychological characteristics, Work load* etc. The necessity of additional information, such as questionnaires, needs to be specified and provided.
- Developing the prototype of the *intelligent soft-computing* tool, which will be able to assess and predict the risk of an air incident or accident by using specific *flexible* rules being able to handle the relationships between particular causes, circumstances and consequences, expressed in both quantitative and qualitative form.
- Testing the *intelligent soft-computing* prototype by using relevant historical data. For such purpose, the historical data about particular air incidents and accidents in both quantitative and qualitative form could be used. The level of risk could be assessed for a planned line or particular flights, air carriers or traffic decisions.
  A simulation of a *real-time* risk assessment has to be carried out, in which the level of risk of a particular flight could be estimated and possible recommendations could be made. They could concern crew or aircraft change or flight delay or cancellation in the case the level of assessed risk is *Unacceptably High*.

The important outcome of the first stage could be the proposed general hierarchical structure of the *flexible* rule-based model of the risk assessment. It may include different groups of factors, such as

- ✓ Crew;
- ✓ Aircraft;
- ✓ Traffic;
- ✓ Weather etc.

For example, for the *Weather* their sub-structure could be:

- ➢ Wind velocity;
- ➢ Temperature;
- ➢ Probability of occurrence of storms and tornadoes etc.

The necessity of additional information, such as questionnaires, could be identified and provided. This could be the base for developing a framework of the prototype of the *intelligent soft-computing* tool.

The stage of testing and simulation could consist of:

- ✓ Carrying out simulation tests and experiments in order to evaluate the analytical and predictive capabilities of the tool.
- ✓ Special simulation tests of a *real-time* risk assessment could be carried out, in which the level of risk of a particular flight could be estimated and possible recommendations to be made. This could include using Internet as a tool for providing *real-time* information and separation of the locations of the decision-making and process (flight) itself.

10.3
*in*TEND: *E*volving *D*istributed *In*telligence System for Evaluation of TENDering

Last decade is characterised with increasing role of the globalisation, including the construction sector. A number of international large-scale construction projects (in road building, modernisation and reconstruction of so called *Trans-European transport corridors*, gas and oil pipelines, crossing the borders of a number of EU-member and associate member states, Trans-national bridges etc.) are at different stage of development.

Majority of subcontractors for such projects are small and medium-size enterprises, which are bidding to win the tender balancing between a bid price that is as "practically high" *as possible* to maximise profit and as "practically low" *as possible* to win the job (McCaffer and Baldwin, 1984; Hegazy et. al., 1994). Their options for a thorough investigation and research, however, are normally, limited by the costs, time and other restrictions.

Additionally, the problem of mark-up estimation and bidding for a tender forecast is a highly unstructured decision problem that is extremely difficult to analyse and formulate an adequate mechanism. Essentially, it is a risk assessment problem under significant uncertainties.

Particularly, for large-scale Trans-national construction projects, the historical data and expertise of consultants from one country are, obviously, not enough to forecast successfully the outcome of a bid. Specifics of each country differ and, thus, a distributed system spanning over the part or the whole of the continent is necessary.

The emerging *e*-technologies, including Internet and *e*-services make possible realisation of such a system, even in *real-time* mode.

Recently, *intelligent* technologies or ***i***-tech are progressively replacing *conventional* techniques for solving this problem, such as regression models, models based on decision analysis etc. (Li, 1996; Li et. al., 1999) All of them, however, lack one important property – they are not adaptive, evolving.

When new information about the tendering process is accumulated, the only option is to re-train such systems from scratch, losing the useful information already extracted from old historical data. This problem has still not full and adequate solution in the wider context of *intelligent* technology (EUNITE, 2000).

An interesting alternative is to build a distributed in several European countries *intelligent* system for evaluation of tendering, specifically for large-scale Trans-national construction projects. This *smart* system would evolve *on-line,* incorporating dynamically the new information collected during the use of the system.

Useful bits of information in form of rules could be exchanged between different nodes, based in different countries electronically in *real-time* mode. The system could be implemented as user-friendly software and could be ported on the Internet. It could function as an *e*-service in *real time* mode.

This distributed *intelligence* system could be considered as a virtual tendering process giving forecasts for the possible outcome of a specific bid based on the tender's data as well as on the ***e***volving **R**ule-based model of the tendering process, build using historical data of similar bids. The proposed system could serve as a prototype of similar systems and have more generic implications to the confluence of *e*-services and the *i*-tech.

10.4 Conclusion

Flexible rule-based models are used as a tool for description of complex and difficult for conventional treatment problems, such as risk assessment in creditworthiness, civil aviation and construction industry. The notion about such application and some illustrative examples has been presented in this chapter.

eR models offer a possibility to modify, generate, up-date in *real time* the knowledge-based systems. A promising direction is their confluence with the e-services such as Internet, mobile communications etc. In this way, the really occurring changes, disturbances and evolution of the processes could be better matched and kept in control. The **ability** of typical rule-based systems **to imitate reasoning** could gain an important property: **to be evolving.**

11 CONCLUSIONS

This book aims to make a step in the direction of setting the basis for understanding, design and analysis of *intelligent* or evolving *flexible* systems. Systems that are able to grow-up in the process of data acquisition starting from no or minimum *'a priori'* information. Systems that are able to adapt to the changes (either external or internal ones) renovating both their structure and parameters *on-line*. Systems that re-use, inherit the positive, useful information from the previous historical cases.

In the same time, systems that are computationally effective (avoiding lengthy and heavy iterative procedures to the maximum possible extent). The ultimate judgement for their efficiency, however, is still their precision, which have to be no worse than that of any other used system.

These systems are transparent (by differ from *black-box* models including their best representative - neural networks) and are able to incorporate any existing expert knowledge (without making this a must, however). In the same time, they make easy a possible extraction of highly comprehensive knowledge in plain English (or in any other preferable human language).

First in the book, the methodological and terminological basis has been established in the Part I. It concerns two main topics, namely:

- ➢ System modelling;
- ➢ *Flexible* models.

Normally used or *conventional* models have been considered at the begining. This includes first principle based models and so-called *black-box* type of models (polynomial, regression, and neural networks). One specific type of neural networks (Radial-Basis Functions) has been mentioned specifically as it represents a kind of a bridge to the *flexible* rule-based models, which are considered next.

In the beginning, some basic principles of the *fuzzy* set theory have been briefly represented. They include both *fuzzy* set definition (a corner stone of the *fuzzy* set theory itself) and the basic operations over *fuzzy* sets (T norms; S-norms; negation; de-fuzzification; degree of similarity).

Flexible models of different types are discussed next:

- ✓ models with *flexible* parameters;
- ✓ models with *flexible* (in)equalities;
- ✓ relational models;
- ✓ Mamdani type models;

- ✓ Takagi-Sugeno (TSK) type models.

It has been argued that the term *flexible* is more appropriate than *fuzzy* in respect to models, systems and control. It has been also mentioned that *fuzzy* sets are, in fact, quite determined in contrast to their name and to the widely accepted impression, particularly in non-specialists. The reason is that after a membership function is determined (a requirement to have a defined *fuzzy* set) the *fuzziness, vagueness* is transformed and mapped to the membership function.

As a consequence of this kind of a paradox, the main difficulty in dealing with *flexible* sets is shifted to the problem of effective and efficient (theoretically, algorithmically, computationally and time-wise) definition and determination of membership functions. In fact *'all is about membership functions'*.

This leads to the problems considered in the Part II of the book: *flexible* models' identification.

The possible approaches to this problem has been divided into two main groups:

- ➢ Non-linear approach;
- ➢ Quasi-linear approach.

The identification problem is formulated with special attention to the identification criteria. A brief introduction to the genetic algorithms (GA) is also given, because they are often used for non-linear identification.

An original crossover operator based on the centre-of-gravity paradigm is introduced. It speeds up convergence without additional computational costs.

Two original approaches to encoding and decoding the FRB model into the chromosome of the GA has been introduced and discussed. Both of them consider two-part chromosome. First part represents the model structure, while the second one encodes the parameters of the membership functions. The two approaches differ by the object of encoding. The first one treats *flexible* rules' indices, while the second concerns linguistic terms' labels.

Comparing to the usually used encoding of all possible rules (normally a huge number) using '0' and '1' both methods are significantly more compact and computationally efficient. Additionally, the second one has higher degree of transparency and provides for exchange of useful informative bits on the level of linguistic terms.

It has been mentioned that the *non-linear* approach to FRB models identification is based on numerical, iterative procedures (gradient-based back-propagation, or, more often, GA, because of their specifics). Although it could provide a global solution (especially GA) and to optimise in the same time both model structure and parameters, including membership functions' shape, type, degree of firing of each rule etc., they are computationally and time expensive and appropriate for ***off-line* applications only**.

The alternative option is to apply *quasi-linear* approach. It concerns TSK models benefiting from their dual nature (being linear from the output and non-linear from the input).

Initially, the structure of the rule-based model is determined **non-iteratively** by data space clustering. Parameters of the consequent part are then estimated by least squares technique (or also by clustering, when singletons are used as outputs) also **non-iteratively**.

Application of more general parameterised de-fuzzification approach did not affect the computational efficiency, because the consequent part is still linear in respect to parameters. It, however, adds to the model *flexibility*.

The formula for informative potentials of each data point, central to the *non-iterative* data space clustering approaches, has been normalised, which is an important generalisation from the point of view of the possible application of the approach in *on-line* mode.

Possibilities for simplification of the resulting *flexible* rule-based model using rules and linguistic terms' similarity have been considered. The option for model parameters' refinement by further non-linear optimisation has been mentioned. Model structure simplification and parameter refinement is closely related to the trade-off between the precision and the general descriptive power of the model.

It has been illustrated on the example of a fan-coil sub-system, that model structure could be simplified significantly (more than two times less rules) without decrease in the precision.

The need for *on-line* identification algorithms has been emphasised. The notion about *intelligent* and *smart* adaptive systems has been presented. A loose definition, problems and specifics as well as the importance of such systems has been mentioned. Practical implications and specific features of *smart* adaptive systems have been illustrated with an example of an *intelligent* Indoor Climate Control (*i*ICC) System.

Chapter 7 is central in the sense that it represents the newly introduced approach for *on-line* *i*dentification of *flexible* models. The concept of the approach is outlined and the basic stages of the procedure are given. The main stages are:

- ✓ potentials up-date in *on-line* mode;
- ✓ rule-base *innovation* mechanism;
- ✓ parameters up-date.

*E*volving **R**ule-based (***e*R**) models has an interesting and promising feature: they are able to build-up and up-grade their structure *on-line* by *learning trough experience* using a minimum amount of *a priori* knowledge. This specific of ***e*R** models is potentially very useful in robotics and process control.

Model structure could be further simplified in *on-line* mode based on rules similarity. Parameters of the FRB model could, optionally, be optimised/refined by a numerical approach like GA. This is, however, iterative and time-consuming procedure, and therefore the recommendation is to be applied at every K steps (K>>1) only.

An interesting application of ***e*R** models is a truly adaptive and evolving control using so called indirect adaptive control scheme.

The proposed methodology has been applied to different *real* engineering problems:

✓ modelling components of HVAC systems;

- heating/cooling coils (outlet air temperature, heat transfer);
- fans (total power and pressure drop across the fan);
- boilers (fuel consumption and boiler efficiency).

✓ modelling the thermal load of a building;
✓ optimal scheduling of a hollow core ventilated slab system;
✓ lactose oxidation fermentation;
✓ creditworthiness assessment;
✓ risk assessment in civil aviation;
✓ evaluation of tendering for large-scale international construction projects.

The concept of *'learning trough experience'* has been validated with *real* experimental data (*on-line* modelling of the valve control of a cooling coil and the cell mass growth) and proved to be effective. ***e*R** models have been successfully applied to modelling dynamical signals also.

Parameters of membership functions describing respective linguistic terms have been optimised by a GA. A twofold improvement of the model precision has place for validation data (for training data the improvements was even higher - three times), but this was for the expense of the loss of the linguistic concept for some of the linguistic terms. Additionally, this model refinement is time consuming and problem-dependent.

The models could be used for control (in which case the precision should be balanced with the computational efficiency), for classification (in which case the precision is paramount and *off-line* mode is perfectly feasible) or for fault detection and diagnostics (in which case the linguistic transparency is highly desired). The frequency of the refinements has to depend on the purpose of the model and specifics of the problem (time constants etc.).

The problem of ICC system modelling and control has been treated from the point of view of the system theory. A specific of these systems is the presence of the human (occupant) which determine in a *fuzzy, vague* way the goal of the system.

This problem has broader implications to the so-called *'consumer-based'* systems, including Internet. It has been argued that the consumer (occupant(s)) has to be considered as a part of the system and the control loop has to be closed through this consumer.

This leads to the necessity of building appropriate re-usable and adaptive approaches for behavioural modelling in *on-line* mode. ***e*R** models, presented in this book, are seen by the author as a promising tool for this purpose.

Their main advantage is that they combine in a specific way the adaptive, evolving capability with the potential to contain and express expert linguistic knowledge together with the quantitative data available in *real-time*. The unique combination of *flexibility* and adaptivity makes them a powerful tool for design of *intelligent* and *smart* adaptive systems.

REFERENCES

Andersen H.C., F.C. Teng, A.C. Tsoi (1994) Single Net Indirect Learning Architecture, *IEEE Transactions on Neural Networks*, **v.5** (6), pp.1003-1005

Angelov P. (2001) Supplementary Crossover operator for Genetic Algorithms, *Control and Cybernetics*, **v.30** (2), pp.1-18

Angelov P. (2000) Evolving Fuzzy Rule-based Models, *Journal of CIIE, **special issue** on Soft Computing Applications to Industrial Engineering*, **v. 17**, pp. 459-468

Angelov P. (1999) A Fuzzy Approach to Building Thermal Systems Optimization, *Proc. of the 8th IFSA World Congress*, Taipei, Taiwan, **v. 2**, pp. 423-426

Angelov P. (1995) An Analytical Method for Solving a Type of Fuzzy Optimization Problems, *Control and Cybernetics*, **v. 24** (3), pp.363-373

Angelov P. (1993) *An Approach to Optimal Control of Biotechnological Processes*, Ph.D. Thesis, Sofia: Bulgarian Academy of Sciences

Angelov P., Simova E., D. Beshkova (1996) Control of Cell Protein Synthesis from Kluyweromyces Marxianus var. Lactis MC5, *Biotecchnology and Biotechnological Equipment* **v.10** (1), pp.44-50

Angelov P., R. Buswell (2001a) Recursive *On-line* Identification of Takagi-Sugeno Models by Rules and Parameters Innovation, *IEEE Transactions on Fuzzy Systems*, 2001, to appear

Angelov P., R. Buswell (2001b) **E**volving **R**ule-based Models: A Tool for *Intelligent* Adaptation, *Joint 9th IFSA World Congress and 20th NAFIPS Annual Conference*, Vancouver, BC, Canada, 25-28 July, pp.1062-1067, invited paper

Angelov P., R. Guthke (1997) A GA-based Approach to Optimization of Bioprocesses Described by Fuzzy Rules, *Journal of Bioprocess Engineering*, **v. 16**, pp.299-301

Angelov P.P., V. I. Hanby, J.A. Wright (2000a) HVAC Systems Simulation: A Self-Structuring Fuzzy Rule-Based Approach, *International Journal of Architectural Sciences*, **v.1** (1), pp.49-58, invited paper

Angelov P.P., V.I. Hanby, R. A. Buswell, J.A. Wright (2000b) Automatic Generation of Fuzzy Rule-based Models from Data by Genetic Algorithms, *In: Developments in Soft Computing* (R. John and R. Birkenhead Eds.): Springer Verlag, pp.31-40

Angelov P.P., V.I. Hanby, R. A. Buswell, J.A. Wright (2000c) A Methodology for Modelling HVAC Components using Evolving Fuzzy Rules, *IEEE International conference on Industrial Engineering, Control and Instrumentation IECON-2000*, 22-28 October 2000, Nagoya, Japan, 247-252, invited paper

Angelov P., D. Lakov (1998) Fuzzy Rule-based System for Risk Assessment, *Proc. of the International Conference on Intelligent Control'98*, Sofia, pp.42-45

Angelov P., S. Tzonkov (1993) Optimal Control of Biotechnological Processes Described by Fuzzy Sets, *Journal of Process Control*, **v.3** (3), pp.147-152

Angelov P.P., J.A. Wright (2000) A Centre-of-gravity-based Recombination Operator for GA, *IEEE Conference IECON-2000*, Nagoya, Japan, 22-28 Oct. 2000, pp.259-264

Astrom K., B. Wittenmark (1984) Computer Controlled Systems: Theory and Design, Englewood Cliffs, NJ, USA: Prentice Hall

Babuska R., H. B. Verbruggen, H. Hellendoorn (1999) Promising Fuzzy Modeling and Control Methodologies for Industrial Applications. *Proc of the European Symposium on Intelligent Technologies ESIT'99*, AB-02, Crete, Greece, http://lcewww.et.tudelft.nl/~babuska/bib/index.html

Babuska R. (1999), Data-driven Fuzzy Modeling: Transparency and Complexity Issues, *Proc. of the Euro Symposium on Intelligent Technologies ESIT'99*, AB-01, Crete, Greece, June 1999 http://lcewww.et.tudelft.nl/~babuska/bib/index.html

Baldwin J., B. Pilsworth (1982), Dynamic Programming for Fuzzy Systems with Fuzzy Environment, *Journal of Mathematical Analysis and Applications*, **v.85**, pp.1-23

Bastian A. (1996) A Genetic Algorithm for Tuning Membership Functions, *Proc. of the 4th European Congress on Fuzzy and Intelligent Technologies EUFIT'96*, Aachen, Germany, **v.1**, pp.494-498

Bellman R., L. Zadeh (1970) Decision Making in a Fuzzy Environment, *Management Science*, **v.17**, pp.141-160

Bentley P. J. (2000a) Evolving Fuzzy Detectives: An Investigation into the Evolutionm of Fuzzy Rules, *In: Suzuki, Roy, Ovasks, Furuhashi and Dote (Eds), Soft Computing in Industrial Applications*, London: Springer Verlag

Bentley P.J. (2000b) 'Evolutionary, my Dear Watson': Investigating Committee-based Evolution of Fuzzy Rules for the Detection of Suspicious Insurance Claims, *Proc. of the 2nd Genetic and Evolutionary Computation Conference (GECCO 2000)*, July 8-12, Las Vegas, Nevada, USA

Bentley, P. J. and Corne, D. W. (Eds.) Creative Evolutionary Systems, San Francisco, CA: Morgan Kaufmann Publishers Inc

Bernard O., G. Bastin, P. Angelov (1999) Hybrid Modelling of Biotechnological Proceses using Neural Networks, *Proc. of the IFAC World Congress*, Beijing, **v.O**, pp.469-474

Bettenhausen K., S. Gehlen, P. Marenbach, H. Tolle (1995) BioX++ - New Results and Conceptions Concerning the Intelligent Control of Biotechnological Processes, *Proc. of the 6th Intern. Conference on Computer Applications in Biotechology*, Garmish-Paterkirchen, Germany, pp. 324-331

Bezdek J. (1974) Cluster Validity with Fuzzy Sets, *Journal of Cybernetics*, **v.3** (3), pp.58-71

Bigus J., J. Bigus (1998) *Constructing Intelligent Agents with Java: A Programmers Guide to Smarter Applications*, Toronto, Canada: John Wiley and Sons Inc.

Brager G.S., R. J. de Dear (1998) Thermal Adaptation in the Build Environment: A Literature Review, *Energy and Buildings*, **v. 27** (1), pp.83-96

Brandemuehl M.J., S. Gabel, I. Andersen (1998) *HVAC 2 Toolkit: A Toolkit for Secondary HVAC System Energy Calculations*, ASHRAE, TC 4.7

Braithwaite G.R., R. E. Caves (1997) Airline Safety-Some Lessons from Australia, *The Aeronautical Journal*, pp. 29-32.

Burkhardt D.G., P.P. Bonissone (1992) Automated Fuzzy Knowledge Base Generation and Tuning, *Proc. of the 1st IEEE Fuzzy Systems Conference*, pp.179-188

Carlsson C., R.Fuller (2001), Optimization under Fuzzy IF-THEN Rules, *Fuzzy Sets and Systems*, **v.119**, pp.111-120

Carse B., T.C. Fogarty, A. Munro (1996) Evolving Fuzzy Rule-based Controllers using GA, *Fuzzy Sets and Systems*, **v.80**, pp.273-294

Castillo L., A. González, and R. Pérez (2001) Including a Simplicity Criterion in the Selection of the Best Rule in a Genetic Fuzzy Learning Algorithm, *Fuzzy Sets and Systems*, **v.120** (2), pp.309-321

Cios K.J., W. Pedricz, R.W. Swinarski (1998) *Data Mining Methods for Knowledge Discovery*, Boston, MA, USA: Kluwer Academic Press

Chen L., O. Bernard, G. Bastin, P. Angelov (2000) Hybrid Modelling of Biotechnological Processes using Neural Networks, *Control Engineering Practice*, **v.8** (7), pp.821-827

Chen L., N. Tokuda, X. Zhang, Y. He (2001) A New Scheme for an Automatic Generation of Multi-variable Fuzzy Systems, *Fuzzy Sets and Systems*, **v.121**, pp.323-329

Chiang C. K., H.-Y. Chung, J.J. Lin (1996), A Self-Learning Fuzzy Logic Controller using Genetic Algorithms with Reinforcements, *IEEE Trans. on Fuzzy Systems*, **v.5**, pp.460-467

Chiu S.L. (1994) Fuzzy Model Identification based on Cluster Estimation, *Journal of Intelligent and Fuzzy Systems*, **v.2**, pp.267-278

Cole J., (1997) Overview of Aviation Safety Issues, *Proc. of the 7th Annual Aviation Forecast Conference*, NATCA -National Air Traffic Controllers Association, Washington DC, USA, p. 6

Corrie S. J. (1994) Potential Growth in Air Travel Demands Renewed Effort to Improve Safety Record, *ICAO Journal*, International Civil Aviation Organisation, Montreal, Canada, pp. 7-9

Cooper M. G., J. J. Vidal (1996) Genetic Design of Fuzzy Controllers, *In: Genetic Algorithms and Pattern Recognition*, S. K. Pal, P. P. Wang Eds., CRC Press, chapter **v.13**, pp.283-298

Davis L. (1989) Adapting Operator Probabilities in Genetic Algorithms, *Proc. of the International Conference on Genetic Algorithms ICGA89*, pp.61-69

Diaz G., M. Sen, K. T. Yang, R. L. McClain (1999) Simulation of Heat Exchanger Performance by Artificial Neural Networks, *HVAC&R Research*, **v.5** (3), pp.195-208

Driankov D., H. Hellendoorn, M. Reinfrank (1993) An Introduction to Fuzzy Control, Berlin, Germany: Springer Verlag

Dubois D., H.T. Nguyen, H. Prade, M. Sugeno (1998) *Fuzzy Systems*, Boston, MA, USA: Kluwer

Eerikainen T., Linko S., Linko P., Siimes T., Zhu Y.-H., (1993). Fuzzy Logic and Neural Network Applications in Food Science and Technology, *Trends in Food Science and Technology*, **v.4**, pp.237-242

Eerikainen T., Zhu Y.-H., Linko P., (1993) An Expert System with Fuzzy Variables and Neural Network Estimation, *1st Euro Congress on Fuzzy and Intel. Tech. EUFIT'93,* Aachen, Germany, **v.1**, pp.202-207

EUNITE (2000) *EUropean Network on Intelligent Technologies for Smart Adaptive Systems*, Contract No IST-2000-29207, Project Summary, p.4

Evans A. W. (1996) Risk Assessment by Transport Organisations, *Transport Review*, **v. 17** (2), pp.145-163

Filev D. (1991) Fuzzy Modelling of Complex Systems, *International Journal of Approximate Reasoning*, **v.5**, pp.281-290

Filev D., P. Angelov (1992) Fuzzy Optimal Control, *Fuzzy Sets and Systems*, **v.48** (2), pp.151-156

Filev D., P. Angelov (1992) Optimal Control in a Fuzzy Environment, *Yugoslav Journal on Operations Research*, **v.2** (1), pp.33-43

Filev D., P. Angelov (1991) Optimal Control under Uncertainties of Fuzzy Type, *Automatics and Informatics*, **v. 9-10**, pp.1-2

Filev D., T. Larson, L. Ma (2000a) Intelligent Control for Automotive Manufacturing - Rule-based Guided Adaptation, *Proc. of the IEEE Conference IECON-2000*, Nagoya, Japan, October 2000, pp.283-288

Filev D., R. Yager (1991) A Generalized Defuzzification Method via BAD Distributions, *International Journal of Intelligent Systems*, **v.6**, pp.687-697

Fedrizzi M., M. Fedrizzi (1993) Equality Evaluation of Public Risk Using Fuzzy Sets, *5th IFSA World Congress*, Seoul, Korea, pp.676-678

Friedman J.H. (1991) Multivariate Adaptive Regression Splines, *The Annals of Statistics*, **v.19** (1), pp.1-141

Furuhashi T., K. Nakaoka, Y. Uchikawa (1995) An Efficient Finding of Fuzzy Rules using a New Approach to Genetic-based Machine Learning, *Proc. of the IEEE Conference on Fuzzy Engineering*, Yokohama, Japan, pp.715-722

Fogarty T.C. (1989) Varying the Probability of Mutation in The Genetic Algorithm, *Proc. of the International Conference on Genetic Algorithms ICGA89*, pp.104-109

Geyer-Schulz A. (1995) *Fuzzy Rule-Based Expert Systems and Genetic Machine Learning*, Studies in Fuziness, **v.3**, Berlin, Germany: Physica Verlag

GECCO (2001) Genetic and Evolutionary Computation Conference (GECCO-2001), URL: http://www.isgec.org/GECCO-2001

Ghiaus C. (2001) Fuzzy Model and Control of a Fan-coil, *Energy and Buildings*, **v.33**, pp.545-551

Goldberg D.E. (1989) *Genetic Algorithms in Search, Optimization and machine Learning*, Reading, MA, USA: Addison-Wesley

Grefenstette J.J. (1986) Optimization of Control Parameters for Genetic Algorithms, *IEEE Transactions. on Systems Man and Cybernetics*, v.**16** (1) pp.122-128

Guthke, R., W. Rausch (1994) Model Aided Multiple Correlation Analysis between Precuture and Main Fed-Batch Culture. *In: Galindo, E; Ramirez O.T.: Advances in Bioprocess Engineering*, pp.267-274. The Netherlands: Kluwer Academic Publishers

Hagras H., V. Callaghan, M. Colley (1999) A Fuzzy-Genetic Based Embedded-Agent Approach to Learning and Control in Agricultural Autonomous Vehicles, *Proc. Of the IEEE Conference On Robotics and Automation*, Detroit, USA, pp.1005-1010

Hanby V. I., J.A. Wright (1989) HVAC Modelling Studies, *Building Services Engineering Research and Technology*, **v.10**, pp.35-39

Hegazy T., O. Moselhi (1994) Analogy–based Solution to Mark-up Estimation Problem, Journal of Computation in Civil Engineering, **v.8** (1) pp.72-87

Hepworth S.J., A.L. Dexter, Willis S.T.P. (1994) Neural Network Control of a Non-linear Heater Battery, *Building Services Engineering Research and Technology*, **v.15** (3), pp. 119-129

Hoffmann F., G. Pfister (1996) Learning of a Fuzzy Control Rule base Using Messy Genetic Algorithms, *In: Herrera and Verdegay (Eds.) Studies in Fuzziness and Soft Computing*, **v.8**, pp.279-305, Heidelberg, Germany: Physica Verlag

Hornik K. (1991) Approximation Capabilities of Multilayer Feedforward Network, *Neural Network*, **v.4**, pp.251-257

Ishibuchi H., Nakashima T., Murata T. (1999) Performance Evaluation of Fuzzy Classifier Systems for Multidimensional Pattern Classification Problems, *IEEE Transactions on SMC-B*, **v.29**, pp.601-618.

Jang J.S.R. (1993) ANFIS: Adaptive Network-based Fuzzy Inference Systems, IEEE Transactions on Systems, Man & Cybernetics, **v.23** (3), pp.665-685

Jang J.-S.R., C.-T. Sun, E. Mizutani (1997) *Neuro-Fuzzy and Soft Computing: A Computational Approach to Learning and Machine Intelligence*, Upper Sadle River, USA: Prentice Hall

Janic M. (2000) An Assessment of Risk and Safety in Civil Aviation, *Journal of Air Transport Management*, **6** (1), 43-50.

Kacprzyk J.(1995) Multistage Control of a Fuzzy System using a Genetic Algorithm,*4th FUZZ-IEEE/IFES Congress*, Yokohama, Japan, pp. 1083-1088

Kacprzyk J. (1983) A Generalization of Fuzzy Multistage Decision Making and Control via Linguistic Quantifiers, *International Journal of Control*, **v.38**, pp.1249-1270

Kacprzyk J., S. Orlovski Eds. (1987) *Optimization Models Using Fuzzy Sets and Possibility Theory*, Dordrecht, Germany

Klir G., T. Folger (1988) *Fuzzy Sets, Uncertainty and Information*, Englewood Cliffs, NJ: Prentice Hall

Kuntze H.-B., T. Bernard (1998) A New Fuzzy-based Supervisory Control Concept for the Demand-responsive Optimization of HVAC Control Systems, *Proc. of the 37th IEEE Conference on Decision and Control*, Tampa, Florida, USA, pp.4258-4263

Kosko B. (1992) *Neural Networks and Fuzzy Systems: A Dynamical Systems Approach to Machine Intelligence*, Englewood Cliffs, NJ, USA: Prentice Hall

Lakov D. (1997) Fuzzy Neural Network Structures in Credit Risk Assessment, *7th IFSA World Congress*, Prague, Czech Republic, **v. IV**, pp. 109-115

Lakov D. (1997) Consecutive Fuzzy Systems in Credit Risk Assessment, *International Workshop on Intelligent Control INCON'97*, Sofia, Bulgaria, pp.73-77

Lai Y. J., C.L.Hwang (1993) Possibilistic Linear Programming for Managing Interest Rate Risk, *Fuzzy Sets and Systems*, **v. 54,** pp.135-146

Lea R., E. Dohmann, W. Prebilsky, Y. Jani (1996) An HVAC Fuzzy Logic Zone Control System and Performance Results, *Proc. IEEE Conference*, pp. 2175-2180

Li H. (1996) Neural Network Models for Intelligent Support of mark-up Estimation, Engineering Construction and Architectural Management, **v.3** (1-2) pp.69-81

Li H., L. Y. Shen, P. E. D. Love (1999) ANN-Based Mark-up Estimation System with Self-Explanatory Capacities, *Journal of Construction Engineering Management*, **v.125** (3), pp.185-189

Lim M.H., S. Rahardja, B.H. Gwee (1996) A GA Paradigm for Learning Fuzzy Rules, *Fuzzy Sets and Systems*, **v.82**, pp.177-186

Linko (1988) Uncertainties, Fuzzy Reasoning and Expert Systems in Bioengineering, *Annals of the New York Academy of Sciences*, **v.542**, pp.83-101

Linko P., T. Eerikainen, S. Linko, Y.-H. Zhu (1993) Artifical Intelligence for the Food Industry, *Proc. of the Conference on AI for Agriculture and Food-Equipment and Process Control AIFA'93*, Paris, France, pp.187-200

Linko P. et al. (1994), Hybrid Fuzzy Neural Bioprocess Control, *Proc. of the 2nd European Congress on Intelligent Techniques and Soft Computing, EUFIT'94*, Aachen, Germany, **v.1**, pp.84-90

Ljung L. (1987) *System Identification: Theory for the User*, New Jersey, USA: Prentice-Hall

Loveday D.L., G.Virk (1992) Artificial Intelligence for Buildings, *Applied Energy*, **v.41**, pp.201-221

McCaffer R., A. Baldwin (1994) *Estimating and Tendering for Civil Engineering Works*, London,UK: Granada Technical Books

Mamdani E.H., Assilian S. (1975) An Experiment in Linguistic Synthesis with a Fuzzy Logic Controller, *International Journal of Man-Machine Studies*, **v.7**, pp.1-13

Mamdani E.H. (1977) Application of Fuzzy Logic to Approximate Reasoning using Linguistic Systems, *Fuzzy Sets and Systems*, **v.26**, pp.1182-1191

Mamdani E.H., Assilian S. (1975) An Experiment in Linguistic Synthesis with a Fuzzy Logic Controller, *International Journal of Man-Machine Studies*, **v.7**, pp.1-13

Marmelstein R. E., G. B. Lamont (1998) Evolving Compact Decision Rule Sets, *In: J. R. Koza (Ed.) Late Braking Papers at the Genetic Programming 1997 Conference*, Omni Press, pp.144-150

Michalewicz Z. (1996) Genetic Algorithms + Data Structures = Evolution Programs, Berlin, Germany: Springer Verlag

Michalewicz Z., D.B. Fogel (1999) How to Solve It: Modern Heuristics, Berlin, Germany: Springer Verlag

Mistry S. I., S.S.Nair (1993) Non-linear HVAC Computations using Neural Networks, *ASHRAE Transactions*, **v.99,** pp.775-784

Muhlenbein H., D. Schlierkamp-Voosen (1993) Predictive Models for the Breeder Genetic Algorithm, *Evolutionary Computation*, **v.1**, pp.25-49.

Mosheli O., T. Hegazy, P. Fazio (1993) DBID: Analogy-based DSS for Bidding in Construction, *Journal of Construction Engineering Management*, **v.119** (3), pp.63-69

Narendra K.S., J. Balakrishnan, M.K.Ciliz (1995) Adaptation and Learning using Multiple Models, switching and tuning, *IEEE Transactions on Control Systems*, pp.37-50

Nelles O. (1996) FUREGA - Fuzzy Rule Extraction by GA, *4th European Congress on Fuzzy and Intelligent Technologies EUFIT'96*, Aachen, Germany, **v.1**, pp.489-493

Nozaki K., T. Morisawa, H. Ishibuchi (1995) Adjusting Membership Functions in Fuzzy Rule-based Classification Systems, *Proc. of the 3d European Congress on Fuzzy and Intelligent Technologies, EUFIT'95*, Aachen, Germany, **v.1**, pp.615-619

Orlovski S. (1978) Decision Making with a Fuzzy Preference Relation, *Fuzzy Sets and Systems*, **v.1,** pp.155-167

Pal S.K., P.P.Wang (1996) *Genetic Algorithms for Pattern recognition,* CRC Press, Boca Raton, FL

Patanik L.M., S.Mandavali (1996) Adaptation in Genetic Algorithms, *In: Genetic Algorithms for Pattern recognition, S. K. Pal, P.P. Wang, Eds.*, CRC Press, Boca Raton, FL, pp.45-64

Pedrycz W. (1984) An Identification Algorithm in Fuzzy Relational Systems, *Fuzzy Sets and Systems*, **v.13**, pp.153-167

Pedrycz W. (1993) *Fuzzy Control and Fuzzy Systems,* (2nd edition) New York, USA: John Wiley and Sons

Ren M.J. (1997) *Optimal Predictive Supervisory Control of Fabric Thermal Storage Systems,* Ph.D. Thesis, Loughborough University, Loughborough, UK

Rivera S. L., M. N. Karim (1993) Use of Micro-Genetic Algorithms in Bioprocess Optimization, *Proc. of 12th World Congress of IFAC*, Sydney, Australia, **v.8**, pp.217-220

Roubos J.A., M. Setnes (2000) Compact Fuzzy Models through Complexity Reduction and Evolutionary Optimization, *Proc. of the IEEE Conference on Fuzzy Systems FUZZ-IEEE*, San Antonio, USA, pp.762-767

Rumelhart D.E., J.L. McClelland, and the PDP Research Group (1986) *Parallel Distributed Processing*, **1**, *Foundations*, Cambridge, MA: MIT Press

Rommelfanger H. (1989) Inequality Relations in Fuzzy Constraints and its use in Linear Fuzzy Optimization, *in: Verdegay J.L., Delgado M. (Eds.): The Interface Between Artificial Intelligence and Operational Research in Fuzzy Environment*, Rheinland, Koln, Germany: Verlag TUV, pp.195-211

Salsbury T.I., R.C.Diamond (2001) Fault Detection in HVAC Systems Using Model-based Feed Forward Control, *Energy and Buildings*, **33**, 403-415

Schwefel, H.-P. (1981) Numerical Optimization of Computer Models, Chichester, NJ, USA: John Wiley and Sons

Setnes M., J.A. Roubos (1999) Transparent Fuzzy Modelling using Clustering and GA's, *Proc. of the NAFIPS Conference*, New York, USA, pp.198-202

Setnes M., J.A. Roubos (2000) GA-Fuzzy Modelling and Classification: Complexity and Performance, *IEEE Transactions on Fuzzy Systems,* **8** (5), 509-522

Seo F. (1999) Multiple Risk Assessment with Possibilistic Utility Models in Incomplete Information Structure, *Proc. 8th IFSA World Congress*, Taiwan, Taipei, 17-20 August, **v.2**, pp.576-580

Shi, Z., K. Shimizu (1992) Neuro-Fuzzy Control of Bioreactor Systems whith Pattern Recognition, *Journal of Fermentation Bioengineering*, **v.74**, pp.39-45

Shimizu K., K. Ye (1995) Development of Intelligent Control Systems for Bioreactor, *Proc. of the 6th Intern. Conf. on Computer Applications in Biotech*, Garmish - Paterkirchen, Germany, pp.89-94.

Shimojima K., T. Fukuda, Y. Hashegawa (1995) Self-Tuning Modeling with Adaptive Membership Function, Rules, and Hierarchical Structure based on Genetic Algorithm, *Fuzzy Sets and Systems*, **v.71**, pp.295-309

So A. T. P., W.L. Chan, W.L. Tse (1994) Fuzzy Air-Handling System Controller, *Building Services Engineering Research and Technology*, v.**15** (2) pp.95-105

Staniskis J., V. Kildishas, D. Filev (1988) *Improvement of the Control of Biotechnological Processes*, Vilnius, Lithuania: LitNIINTI (in Russian)

Strobach P. (1990) Linear Prediction Theory: A Mathematical Basis for Adaptive Systems, New York, USA: Springer Verlag

Su H.-T., T.J. McAvoj, P.Werbos (1992) Long-term Predictions of Chemical Processes using Recurrent Neural Networks: a Parallel Training Approach, *Industrial Engineering and Chemical Research*, **v.31**, pp.1338-1352

Sugeno M., T. Yasukawa (1993) A Fuzzy-Logic-Based Approach to Qualitative Modeling, *IEEE Transactions on Fuzzy Systems*, **v.1** (1), pp.7-31

Takagi H., M. Lee (1994) Neural Networks and Genetic Algorithm Approaches to Auto-Design of Fuzzy Systems, *In: Lecture Notes on Computer Science, Proc. of FLAI'93*, Linz, Austria: Springer Verlag, pp.68-79

Takagi T., M. Sugeno (1985) Fuzzy Identification of Systems and its Application to Modelling and Control, *IEEE Transactions on Systems, Man and Cybernetics*, **v.15**, pp.116-132

Tanaka H., Asai K. (1984) Fuzzy Linear Programming Problems with Fuzzy Numbers, *Fuzzy Sets and Systems*, **v.13**, pp.1-10

Taylor F. (1995) *Moderate Thermal Environments - Determination of the PMV and PPD indices and specification of the conditions for thermal comfort, BS EH ISO 7730*: ISO

Turksen I.B., I.A. Willson (1994) A Fuzzy set Preference Model for Consumer Choice, *Fuzzy Sets and Systems*, **v.68,** pp.252-266

Valente de Olivieira J. (1999) Semantic Constraints for Membership Function Optimisation, *IEEE Transactions on Systems, Man and Cybernetics-B*, **v.29**, pp.128-138

van den Brink T., M. Albers, B. Zuidwijk (1995) Fuzzy Credit Risk Assessment, *3^d^ Euro Congress on Inteligent Technologies and Soft Computing EUFIT'95*, Aachen, Germany, **v.2**, pp.1704-1709

Wallrafen J., P. Protzel, H. Popp, J. Baetge (1995) Bankruptcy Prediction Using Different Soft Computing Methods, *3^d^ European Congress on Intelligent*

Technologies and Soft Computing EUFIT'95, Aachen, Germany, **v.2,** pp.1710-1714

Wang L.-X. (1992) Fuzzy Systems are Universal Approximators, *Proc. of the International Conference on Fuzzy Systems*, San Diego, CA, USA, pp.1163-1170

Werbos P. (1990) Backpropagation Trough Time: What it Does and How to do it, *Proc. of the IEEE Conference on Neural Networks*, **v.78** (10), pp.1550-1560

Wright J. A. (1991) HVAC Optimisation Studies: Steady-state Fan Model, *Building Services Engineering Research and Technology*, **v.12** (4), pp.129-135

Yager R.R., D.P. Filev (1993) Learning of Fuzzy Rules by Mountain Clustering, *Proc. of SPIE Conf. on Application of Fuzzy Logic Technology*, Boston, MA, USA, pp.246-254

Yager R., D. Filev (1994) *Essentials of Fuzzy Modeling and Control*, NewYork, USA: John Wiley and Sons

Yager R., L. S. Goldstein, E. Mendels (1994) FUZMAR: An Approach to Aggregating Market Research Data Based on Fuzzy Reasoning, *Fuzzy Sets and Systems*, **v.58**, pp.343-354

Zadeh L. A. (1965) Fuzzy Sets, *Information and Control*, **v.8**, pp.338-353

Zadeh L. A. (1973) Outline of a New Approach to Analysis of Complex Systems and Decision Processes, *IEEE Transactions on Systems, Man and Cybernetics*, **v.1**, pp.28-44

Zadeh L. A. (1993) Soft Computing, *Introductory Lecture for the 1st European Congress on Fuzzy and Intelligent Technologies EUFIT'93*, Aachen, Germany, p.vi-vii

Zimmermann H.-J. (1983) Fuzzy Mathematical Programming, *Computers and Operations Research*, **v.10**, 291-298

INDEX

Druck: Strauss Offsetdruck, Mörlenbach
Verarbeitung: Schäffer, Grünstadt

Lightning Source UK Ltd.
Milton Keynes UK
27 April 2010

153419UK00004B/25/A